Sustainable Water Management and Wetland Restoration Strategies in Northern China

Edited by
Giuseppe Tommaso Cirella
Stefan Zerbe

peer-reviewed

On behalf of
Stifterverband für die Deutsche Wissenschaft
Kurt-Eberhard-Bode-Stiftung
für medizinische und naturwissenschaftliche Forschung

Contents

Contents

Foreword

On 26 June 2013, the Water Research Horizon Conference was held in Berlin, together with the Awards Ceremony of, the Kurt-Eberhard-Bode Foundation and Rüdiger-Kurt-Bode Foundation, two foundations supporting research on water and its use in a broader sense. These two foundations have spurred from Bode Chemie GmbH, one of the leading chemical manufacturers in clinical environmental use with specialisation in the field of disinfection and hygienic materials. In 1924, the company was founded by Dr. Kurt Bode in which currently is a part of the Paul Hartmann Group. The company has consistently maintained strong interest in the development and use of scientific research and relating results. In its commitment to this initiative, since 2011, the company also has run the Bode Science Center. As an example, scientific developmental application of new chemical products for hospital use include Bacillol® and Sterillium®, two alcohol-based disinfectants.

In 1987, Eberhard Bode, head of the company for many years and son of the founder Kurt Bode, decided to start the Kurt-Eberhard-Bode Foundation in honour of his father's accomplishments. Historically, the foundation has actively supported scientific research within the fields of information technology, biology, bioinformatics, physics, chemistry and medicine. However, one of Eberhard Bode's convictions has been that interdisciplinary science, well beyond the scope of his own company, needed special attention due to its neglect in place of classical funding instruments concentrated primarily within traditional disciplines. For this reason, the foundation has supported many projects and research groups within fields relating to medical applications, medical applications of new materials, bioinformatics and nutrition-based research.

In 2009, the foundation expanded its scope by emphasising the subject matter of "water and sustainable usage of resources" with a principal goal of strengthening water resource research for globally-oriented sustainable development. This decision prompted the foundation to sponsor a number of junior research groups in the development of integrative models for the sustainable usage of such resources from an interdisciplinary and applied-scientific standpoint. The initial group's focus, "Sustainable water management and wetland restoration in settlements of continental-arid Central Asia (SuWaRest)", was a joint-cooperation between the Free University of Bozen-Bolzano and the University of Greifswald headed by Prof. Dr. Stefan Zerbe and Dr. Niels Thevs. At the Berlin meeting, a second sponsorship funding period, 2013-2015, was awarded to another junior research group within the area of biodiversity in aquatic ecosystems. The second foundation, Rüdiger-Kurt-Bode Foundation, was established in 2009 by Rüdiger Bode, son of Eberhard Bode, a pharmacist and entrepreneur. The Rüdiger-Kurt-Bode Foundation promotes interdisciplinary research within the fields of life and natural sciences. The foundation's board of trustees decided to foster the notion for strategies and concepts for the sustainable utilisation of water resources through interdisciplinary approaches and practical applications. The focus of both foundations on interdisciplinary environmental sciences have also been supported by the Stifterverband für die Deutsche Wissenschaft, in particular, the responsible officer and member of the board, Dr. Marilen Macher. She jointly assisted, with the board, in developing the ideas for funding the particulars for much of the water-related research. In addition, the Deutsche Forschungsgemeinschaft (DFG) and in particular Ute Weber, supported the foundations in the context of identifying reviewers.

The project SuWaRest followed an innovative approach by integrating the natural sciences and environmental engineering with environmental ethics in order to work out potential best-use scenarios for the extremely important but often scarce vital resources humankind depend upon. Water, in this case, is essential and exemplar for our existence. The research consortium of SuWaRest did implement a strong interdisciplinary and international cooperation as well as promoted young scientists from Germany, Italy and China on their way to highly qualified researchers in their fields, i.e. landscape ecology, microbiology, environmental physics, environmental economics and environmental ethics. It is this fundamental notion illustrated throughout the research reported in this book that both the Kurt-Eberhard-Bode Foundation and Rüdiger-Kurt-Bode Foundation will continue to support such solutions in the future.

Munich, 21 July 2014

Arndt Bode (on behalf of the Kurt-Eberhard-Bode Foundation and Rüdiger-Kurt-Bode Foundation)

Preface

Resource water is a top global environmental agenda. Quantity as well as quality of water has become a problem throughout the world. In particular, in arid and semi-arid as well as in fast growing urban-industrial environments societies face over-utilisation and subsequent decrease of water bodies, water pollution as well as very limited access to drinking water. With our international and interdisciplinary project "Sustainable water management and wetland restoration in settlements of continental-arid Central Asia" (SuWaRest), we address this issue in northern China. China's economy and population has been most rapidly growing in the past decade, with the consequence of high pressure on the natural resources. In order to feed this growing number of inhabitants, irrigation agriculture has spread in the arid and semi-arid regions of northern and north western China. Traditional land uses have been changed to intensively managed fields and related food industries. In order to find solutions for a sustainable management of the resource water, on the basis of scientific results, it is necessary to integrate various disciplines into the research. Thus, our project bridges landscape ecology, technical physics, environmental microbiology, socio-economics and environmental ethics.

Our book is a compilation of the research from these differing disciplines and the experience of interlinking land-use adaptation, the sustainable use of water, the restoration of degraded ecosystems and sustainability based-thinking. Within our international research team, based out of Italy, Germany and China, we applied a broad range of methodology to derive comprehensive recommendations for water management and wetland restoration in northern China. On principle, we thrive for general approaches, which also can be applied in similar environments throughout Central Asia and, in particular, settlements. SuWaRest has been funded by the Kurt Eberhard Bode Foundation within the Stifterverband für die Deutsche Wissenschaft, Germany. Their support, has sponsored young researchers, postdoctoral and Ph.D. students, supervised by an international team, in building this interdisciplinary problem solution strategy.

Appreciation is extended to all our colleagues and friends in China, especially local support in Gansu Province and Inner Mongolia Autonomous Region. We believe the research and investment in the SuWaRest project was well worth it, and that the learning-base for our young researchers will build confident and career-oriented researchers in-and-out of academia. The structure of the chapters are framed within the interdisciplines and overlap accordingly. We hope that our findings could pave the way to a more sustainable development of arid and semi-arid regions in Central Asia.

Giuseppe Tommaso Cirella – Free University of Bozen·Bolzano
Stefan Zerbe – Free University of Bozen·Bolzano

Introduction

Stefan Zerbe
Giuseppe Tommaso Cirella
Niels Thevs

1. Water resources: On top of the world's global change agenda

Water is one of the main natural resources. Without water, no life could exist on Earth. Water is also an issue of global change and related to environmental problems and human health. For example, 13 % of humanity, globally, do not have access to clean drinking water and about 3.5 million deaths are related to inadequate water supply and sanitation annually (UNESCO 2012). Due to the over-utilisation of water for agriculture, industry and in the large urban agglomerations, surface water as well as groundwater in many regions of the world has decreased in quantity. It is projected that climatic changes will enhance extreme water related events like floods and droughts (UNESCO 2009). About 25 % of all terrestrial ecosystems are degraded with drylands being the most vulnerable ones (FAO 2011). The most common degradation processes in drylands are destruction of vegetation, soil salinisation and soil erosion. A major driver for those degradation processes is water shortage. Central Asia is one of the world's largest dryland regions. There, inadequate water resource management has resulted in large-scale ecosystem degradation. The desiccation of the Aral Sea and degradation of adjacent terrestrial ecosystems are a well-known example.

Central Asia has an extremely continental-arid climate and is largely covered by deserts, semi-deserts, steppes and alpine ecosystems. In arid environments, agriculture and settlement development depends on large-scale irrigation. Water sources are melting water from glaciers and snow as well as precipitation in the high mountain areas that is Tianshan, Qilianshan, Kunlun, Pamir and Hindu Kush, which often are only temporarily transported by rivers to mostly peripheral desert areas. Large river systems in Central Asia include the Syr Darya, Amu Darya, Ili and Tarim (Giese et al. 1998; Kuzmina & Treshkin 1997; Treshkin 2001). There, the irrigated water is diverted from the rivers and lakes to the surrounding settlements with their agricultural lands.

In recent decades, water increasingly is taken from groundwater reservoirs (e.g. Song et al. (2000) for the Tarim River, northwest China) as the main water source for the natural as well as the anthropogenic ecosystems (Thevs 2007). In particular, growing settlement populations and subsequent agricultural food production in Central Asia (Zerbe et al. 2006) are responsible for this transposition. Growing populations have an increasing demand for food, raw materials and energy. Agriculture is one of the main drivers in Central Asia for land development; however, agriculture is also one of the main drivers for environmental problems, such as decreasing resource water and water pollution – via fertiliser and pesticides. In the past decades, in particular cotton plantations have been increasingly introduced throughout Central Asia. For example, 11 % of the world's cotton production comes from the arid and semi-arid regions of Xinjiang (USDA 2012). The irrigation of cotton fields has led to severe water shortages in those regions. The change of the Aral Sea from an "inland ocean" into a salt desert in only two decades is reflective of this (Breckle et al. 2012).

Currently, also the glaciers throughout the bordering mountains are in retreat due to climatic changes (Unger-Shayesteh et al. 2013). Thus, the river run-off from glacier melt has increased during the last two decades (NDRC 2006). It is generally expected that, due to climate change, the river run-off will decrease and become more variable, once the glaciers have shrunken (Hu et al. 1994; Jiang et al. 1997; Giese & Moßig 2004; Barnett et al. 2005). Therefore, in such a continental-arid climate and against the background of climatic changes water must be used efficiently, in particular for agriculture and in the urban-industrial environments.

Another issue of global water change is pollution (UNESCO 2012). Pollution strongly affects the quality of river water and river sediments (Borin et al. 2009). The latter behaves as a sink of pollution, especially for heavy metals and aromatic compounds. Sediments in rivers may be transported to the sea, spread over riverbanks and tidal marshes or managed, that is actively dredged and disposed of on land. Once deposited on tidal marshes or alluvial areas, the polluted sediments may enter semi-terrestrial ecosystems or agro-ecosystems and may pose a risk by accumulating along the food chain and finally end up in the human body. In order to prevent this, phytoremediation through plants is considered as a management option to clean both water and flooding sediments (Pilon-Smits 2005). Thus, pollutants are removed in an early stage of the food chain. In particular, phytoremediation with reed (*Phragmites australis*) is considered an effective technique to clean up waters and sediments from organic and inorganic pollutants. The release of industrial and municipal waste products in freshwater ecosystems has been become a dramatic issue for the condition of the environment and for human health (Taisan 2009).

2. Wetland and river restoration: A task for the 21st Century

In recent decades, the restoration of degraded and damaged ecosystems has become a challenge for landscape management, nature conservation and sustainable land-use development throughout the world (Bradshaw & Chadwick 1980; Jordan et al. 1987; Urbanska et al. 1997; Perrow & Davy 2002; Temperton & Kirr 2004; Lüderitz & Jüpner 2009; van der Zanden & Cook 2011; Hampicke 2009; Zerbe & Wiegleb 2009). Already in 1995, Daily (1995) stated that about 45 % of the terrestrial land surface has a reduced land-use capacity due to a history of unsustainable management. The restoration of floodplains and related settlement ecosystems in arid and semi-arid environments becomes especially difficult with the increasing limitation of water resources. As rivers and their floodplains provide many ecosystem services, for example the purification of water, combating desertification, the accumulation of carbon, production of biomass and providing habitats for plants and animals, there is a particular focus on restoration (Mant & Janes 2006; Lüderitz & Jüpner 2009). Sound restoration science and practice integrates a multiplicity of disciplines. Additionally, it supports decision-makers towards a better assessment, via integrative components of local community knowledge, stakeholders and scientific tools. The task for the 21st Century will be to harmonise these components, at a local level and the augmentation of educative processes, so wetlands and river systems alike can be promptly recognised as central to a productive natural ecosystem.

3. Alternative land-use systems: Reed as a multi-service species

Against the background of increasing agriculture (e.g. crop cultivation, pastures) in Central Asia and, in particular, unsustainable land-use with subsequent negative consequences of the natural resources, alternative land-use systems have to be taken into account. Drivers for these alternative land-use systems can be key species, which are indigenous and adapted to the local and regional environments. Those key species, trees, shrubs as well as herb or grass species, might provide a bundle of ecosystem services. This has been proven for a number of plant species throughout the world. In central and southern Europe, for example, the tree *Castanea sativa* Mill. is such a multi-service species, providing timber, fruits, honey and also cultural services (Conedera et al. 2004). Sustainably managed, it also provides a habitat for a vast number of organisms and serves for environmental education and recreation. In Central Asia, the genus *Apocynum* has been pointed out by Thevs et al. (2012) as a multi-functional species by, in particular, stressing the possible alternative to non-sustainable cotton production in those arid regions.

Reed (*P. australis*), occurring as cosmopolitan species throughout the world, is also considered a promising species which could serve many and diverse demands for the society (Köbbing et al. 2013). It can play an important role within a water-saving and resource-efficient sustainable land-use strategy. Reed can grow on a great variety of sites with regard to different ground water levels, water level changes in floodplains, nutrient availability, and salinity (Thevs 2007; Lambertini et al. 2008; Haslam 2010). Using its ability for phytoremediation, reed can be placed downstream of irrigated land or settlements so that it can use the drainage water and work on the self-purification of wastewater. Thus, by drain-water utilisation or wastewater treatment, reed will produce biomass, which can be used for a neutral CO_2 energy production and which accumulates carbon in the belowground biomass and organic layer, respectively (Kerschbaumer et al. 2014). Furthermore, reed can grow on sites along the lower reaches of rivers, which only receive water during the summer floods of some years. Such sites pose problems for irrigation because a dry period during spring and early summer, e.g. as encountered at the Tarim River in recent years, results in crop failure. Reed can deliver rather stable biomass yields because it takes up water from the groundwater and thus survives periods during which the river, i.e. surface water, is dry. Reed is used as pasture and fodder as well as it is harvested, in order to use the biomass as raw material for paper production and construction materials. If reed is used as pastureland or as fodder plant, it is grazed or harvested during summer, which might result in a certain export of nutrients. For the biomass utilisation as raw material, reed is harvested during winter. The harvested reed biomass is used as raw material for paper production and for the production of chipboards (Hansmann 2008a). As a natural plant, perfectly adapted to the environment, reed does not require treatment like irrigation, seeding, weeding or herbicide or pesticide treatment. Therefore, *P. australis* offers a huge potential as a valuable resource for rural people especially in developing countries. The potentially available reed biomass is lacking reliable data and it is difficult to quantify because as a natural plant it is not part of official statistics such as agricultural crops and residues.

In the recent past, increasing research has been carried out on assessing the advantages of the energy valorisation of biomass (Mckentry 2002; Koziński et al. 1996; Friberg & Blasiak 2002). Some of those studies have analysed biomass combustion technologies or co-firing options with coal. In addition, reed has been considered as an energy source in northern Europe (Graneli

1984). Also in Spain, it is known that some farms could be dedicated to cultivate plant varieties for bioenergy use, such as Giant Cane (*Arundo donax* L.). In the combustion research, an interesting option is the domestic heating by means of conventional boilers, whose efficiencies using reed as fuel has been also compared with the one obtained with traditional fuel, e.g. wood logs, briquettes and pellets.

4. The culture of water

Water is not valued the same in different cultures. There is a strong interlinkage between water and culture. Religion, ethnic, education, environmental laws, etc. influence the individual behaviour towards as well as the society's management of the resource water. Waste of water and water pollution is often related to a low cultural value of water as one of the most important natural resources. Water has always played a more or less prominent role in cultures, depending on the environmental conditions people had to face. Indigenous cultures are still renowned for their ingenious and sustainable water practices. For example, the Karez well system under the extreme continental-arid climate of northwest China was built by the indigenous people to divert water from the mountains in underground channels to the oasis as a perfect adaptation to this dry environment (UNESCO 2014). Modern practices, however, have often disturbed and overruled these traditional practices, with undesired consequences such as the loss of water by evaporation (Schelwald-van der Kley & Reijerkerk 2014). As an example, the water reservoirs along the Tarim River in China evaporate between 0.1 and 0.21 km^3 per year, which is 10 to 25 % of the water used for irrigation along the Tarim River (Thevs et al. 2015). Moreover, according to Schelwald-van der Kley & Reijerkerk (2014), a cultural impact assessment can be a useful tool in assessing and subsequently mitigating adverse cultural impacts of innovations and developments. They state some important points that incorporate a review process in which predictive methods are used in whether or how an action should proceed, monitoring the impacts that occur and acting on the results of such a monitor-based process.

Within environmental ethics, the concept of culture entails perceptions, values, attitudes and institutions. Culture has a long record in history and has undergone different changes. This is also true for "water cultures". Culture shapes the way that society copes with environmental problems. It also shapes the relation between Western scientists and local communities. Success of conservation objectives must therefore properly address the background of culture. This is true especially if different cultures merge at specific sites, such as in Central Asia. The general concept of sustainability has been addressed by Ott and Döring (2008) and Döring (2009). In a more recent study, the role of water culture has been emphasised from different perspectives, e.g. by Balaji et al. (2009). Concepts of social research in developing countries with respect to nature and environmental conservation have been designed in a large scope of literature. One can choose between different tool kits according to specific local circumstances. One step is to reflect upon the problem, which looks at tools and their appropriateness to the study area.

Most experts will agree that water culture needs some sort of comprehensive approach that takes into account past, present and future use. The use of technical solutions in coordination with integrative social-environmental goals, from a comprehensive perspective, is an urgent matter. The culture of water, from a critical analysis of ecocentrism leads us to move away from such an ecocentric stance, and instead, take a visionary look at strong sustainability as a more encouraging, or alternative, path for water use and its maintainability (Kerschbaumer & Ott 2013). Firmly founded, as a result, environmental improvements are grounded in environmental ethics and the relationship connecting value, or the understanding of worth that keeps human beings in equilibrium with the environment. As water is one of the fundamental needs for human survival, water culture via its use and reuse is an exemplar indicator, and starting point, to develop an equilibrium between society and the resource water.

5. Interdisciplinary research on environmental problems in northern China: The SuWaRest consortium

The overall goal of our international and interdisciplinary research consortium was to provide the scientific basis for a comprehensive water management in settlements and on agricultural lands in Central Asia by considering aspects of the natural as well as social sciences. We investigated settlements and irrigation agriculture in northern China with case studies in the Gansu Province and Inner Mongolia Autonomous Region (Inner Mongolia). One of our focal species which we investigated with this multi-disciplinary approach was reed (*P. australis*). As already stated above it occurs along rivers, at lakes and on wetlands in general and is widespread in our investigation areas in the northern Chinese settlements. Reed is utilised there, in particular, as raw material for paper. Additionally, we focused on its potential to be used as biomass for energy production, in particular at the local level. Despite the fact that animal herding on reed stands is a traditional land use all over Central Asia, which is an important income source and an important feature of the various cultures and ethnics (e.g. Turk and Mongolian people), we hypothesised that reed might offer alternative utilisation options in this region. Additionally, ecosystem services such as, e.g. water purification, nutrient extraction and sand fixation in deserts and steppes, it may also provide income opportunities during winter when other agricultural activities are reduced. Moreover, the perception of water by the local stakeholders was analysed which marks the first study of this kind in continental-arid northwest China.

Within this holistic view on water in settlements and irrigation agriculture, we thus address the following issues:

- environmental ethics and arguments in favour of a strong sustainability viewpoint in Chapter 1;
- describe, in detail, our two investigation areas in Chapter 2, i.e. the Heihe River Basin and Wuliangsuhai Lake with the Hetao Irrigation District;
- the use of microbial communities as bioindicators for land-use systems is outlined in Chapter 3;

- water allocation along the Heihe River and Tugai forest conservation in Ejina is described in Chapter 4;
- the physiology of reed (*P. australis*) with regard to nutrient content and allocation is analysed in Chapter 5 by comparing aquatic and terrestrial habitats;
- with regard to our focal plant species reed, diversity and the role of rhizobacteria associated to *P. australis* is investigated in Chapter 6;
- with regard to reed as a renewable energy source, an insight into possible conversion pathways is given in Chapter 7;
- the economic perspective is shown in Chapter 8 by investigating the livelihood and economy of reed-dominated wetlands; and
- the subject of sustainability is addressed with an index of sustainable functionality derived from data on the Urat Front Banner in Chapter 9, and a comparative look at alternative development scenarios on the Hetao Irrigation District and Wuliangsuhai Lake in Chapter 10.

In a final summary, Chapter 11 discusses our findings against the background of our multi-disciplinary approach and derives management recommendations for waters and wetlands in northern China.

Key references

Bradshaw, A.D. & Chadwick, M.J. 1980. The restoration of land. University of California Press, Berkeley, Calif.

Daily, G.C. 1995. Restoring Value to the World's Degraded Lands. Science, 269(5222): 350–354.

Döring, R. 2009. Natural capital – What's the difference? In: R. Döring ed. Sustainability, natural capital and nature conservation. Marburg: Metropolis-Verlag, 123–142.

Köbbing, J.F., Thevs, N. & Zerbe. 2013. The utilisation of Reed (*Phragmites australis*) – A review. Mires and Peat, 13: 1–14.

Mant, J. & Janes, M. 2006. Restoration of rivers and floodplains. In: J. van Andel & J. Aronson eds. Restoration ecology: The new frontier. Malden, Mass: Blackwell Science, 141–157.

Ott, K. & Döring, R. 2008. Theorie und Praxis starker Nachhaltigkeit. Metropolis-Verlag.

Zerbe, S. & Wiegleb, G. 2009. Renaturierung von Ökosystemen in Mitteleuropa. Spektrum Akademischer Verlag, Heidelberg.

1. SuWaRest, the "Third Culture" and environmental ethics

Konrad Ott

1.1 Introduction on epistemology

The "Sustainable water management and wetland restoration in settlements of continental arid Central Asia" (SuWaRest) project has been a so-called "Mode-II" project in transdisciplinary environmental science. While "Mode-I" projects perform normal disciplinary science, Mode-II projects transgress the boundaries between the natural sciences, technological disciplines and humanities – spanning from cultural studies to economics and even to ethics. Mode-II projects have found some attention in the epistemology of sustainability science (Ziegler & Ott 2011). Given the famous distinction of the "two cultures" of natural sciences on the one hand and humanities on the other (Snow 1959, 1990), Mode-II projects are constitutive of a third epistemic culture that tries to combine the empirical rigor (as in biochemical data mining, Chapter 4 and 5) and technological outcomes of the sciences (as in combustion research, Chapter 7) with economic cost-benefit analysis (as in our studies on reed utilisation, Chapter 8) and with the conceptual and reflexive competencies of the humanities (as in scenario writing, Chapter 10). Therefore, SuWaRest has been a paradigm case of such still emerging "third culture" of transdisciplinary environmental research. As editors and authors hope for, this book illustrates the spirit of such third culture. This chapter presents an argument, why and how this third culture should be grounded in ethics. This argument had been outlined years ago with respect to ecological science within the book, *Ipso Facto*, Ch. 8. (Ott 1997). The experience of the SuWaRest project made me even more convinced that this type of argument holds for the "third culture" in general. The context of discovery of this argument was SuWaRest, but its context of application and justification transcends SuWaRest by far.

With respect to empirical findings ("data") and causal explanations ("natural laws"), natural sciences are value-free. Such findings and explanations should not be biased by the many values scientists may be committed to as moral persons. In any case, science should not be submitted to standards of political correctness. This meaning of science as being value-free intrinsically belongs to the general ethos of science (Ott 1997, Ch. 3 & 5). This meaning, however, does not imply that the overall enterprise of doing science and performing human studies within different epistemic disciplines should be completely devoid of ethics. Some highly general principles are underlying the very practice of epistemic disciplines. In philosophy, types of epistemic knowledge have been connected to some general interests that motivate epistemic research. Jürgen Habermas, for instance, has argued in the line of Max Scheler that different epistemic disciplines serve some practical interest of knowledge, written in German as "Erkenntnisinteresse" (Habermas 1965). Very broadly, the interest of knowledge in the natural sciences is, for example, about taxonomy, explanation, technological control and utilisation of nature as being conceived as "neutral objectivity". The interest of knowledge within the humanities is about orientation within a highly complex social order that is shaped by many cultural, religious and aesthetic traditions and by institutional regulations (e.g. law and economics). The humanities both broaden and deepen the sense for the varieties of human cultures and how the many different modes of being human have been realised throughout history (as in "older civilisations" as in China and Europe). According to Habermas, there are also some other "critical" disciplines (as psychoanalysis, social philosophy and ethics) which serve an interest in individual and political liberation since they bring about a general critical and reflexive attitude towards conventions, ideologies and doctrines which may repress and dominate people. Control over nature, orientation in culture and liberation from ideologies and illusions are general practical human interests, which are served by different epistemic cultures. If so, sciences and humanities are oriented both at (1) propositional and theoretical truth and (2) modes of practice by which human life is organised. Dealing with nature is one basic mode of practice.

If one credits the idea that epistemic disciplines are rooted in such general practical interests of knowledge and if one contributes by one's own research to transdisciplinary environmental research ("third culture"), as our group did in the SuWaRest project, one might, on epistemological reflection, ask which specific kind of interests might be implied in such third culture. What kind of practical interest, if any, might be constitutive to the many research programs in the field of climate change, forestry, agriculture, biodiversity, fisheries, conservation biology, restoration ecology and freshwater allocation. According to Habermas' original conception, all these disciplines must be primarily oriented at technological control and domination over nature. However, this conception is too narrow and it does not fit well within environmental disciplines.

Many researchers involved in this new epistemic culture orient themselves at some broad motives to protect, preserve and even restore parts of nature in the face of over-utilisation, pollution, depletion, degradation and the like. Such protective and restorative motives cannot be simply subsumed under Habermas' triadic structure but require another genuine practical interest of knowledge, which cannot be reduced to technological control over nature. Since environmental disciplines emerged since the 1970s, the original conception of "Erkenntnis und Interesse" must be broadened. There is no argument given by Habermas that there are exactly three practical interests of knowledge irrespectively how epistemic cultures may develop.

Given this argument so far, the general interest of knowledge that underlies transdisciplinary environmental research can, in principle, be combined with ideas, leitmotifs and visions in contemporary environmentalism. Given the spectrum of such ideas ("ecological integrity", "flourishing of life", "healing and saving the Earth", "making peace with nature", "overcoming anthropocentrism", "anthropocene with a human face" and the like) it seems prudent within a scientifically shaped culture to adopt the rather modest idea of sustainability which also has some foothold in global environmental policy-making, even if as lip-service only. If the third culture of environmental sciences is to be grounded in some practical interest of knowledge, and if researches in the field are often motivated by protective objectives, and if there is a spectrum of environmental ideas and vision, and if one favours moral parsimony within the ethics of science, the idea of sustainability shows up as an attractive idea for grounding the third culture ethically.

Such grounding of environmental sciences as "third culture" does not impair the methods and standards of sound science that remain intact. Neither does it impair the freedom of research. The adoption of such underlying interest in sustainability does neither change scientific methods and standards nor does the rejection of such underlying ideas improve one's scientific research with respect to truth. It rather orients scientists on a higher layer of reflection.[1] If so, there is a pathway of ethical reflection that originates from the practice of transdisciplinary environmental research and ultimately immerses into the sources of ethics that are constitutive to the idea of sustainability. The idea of sustainability is clearly an ethical one. Its moral sources stem both from a theory of inter- and intragenerational justice and the overall discourse in environmental ethics (Ott & Döring 2008; Ott 2014). The specific concepts of the general idea of sustainability (weak, intermediate and strong sustainability, see final section of Chapter 10) ultimately depend on these sources.

Given the argument so far, there is an intrinsic logic of reflection starting from the performance of transdisciplinary environmental research, as within the SuWaRest project, and ending in the moral sources within the idea of sustainability. This reflective logic is inescapable from a philosophical perspective even if scientists may, for whatever reasons, prefer to abstract it away from their research. Since most scientists are trained to shy away from

---

1 Such grounding is critical against environmental science as it is restricted to mere data mining, which is, perfectly "objective" but disconnected to any meaningful practical purpose. Such "data positivism" is widespread in the Chinese environmental scientific community.

ethics, they often feel unfamiliar with and uncomfortable in theses ethical realms and prefer to get rid of them by abstracting them away. This escape route of abstraction always remains open to scientists. The SuWaRest project, however, might serve as an example that environmental research may succeed scientifically without abstracting away ethical ideas. In the following chapter, only one out of two moral sources of sustainability will be outlined, namely environmental ethics. The topic of intergenerational justice has been addressed recently elsewhere (Ott 2014).

1.2 Environmental ethics

No concept of nature should contradict scientific insights, but any concept should be open for different cultural interpretations of nature. Values and norms cannot logically be deduced from nature as nature is conceived scientifically as value-free objectivity, because from a given set of empirical-descriptive statements, it cannot be deduced what should be done. This would be a so-called naturalistic fallacy. The third culture is grounded in ethics but such grounding does not rest on such fallacy. The argument being presented in this chapter does not derive values or rules from nature itself but it grounds them as presuppositions being implied in the epistemic practices of the third culture. Therefore, the argument belongs to a type of "transcendental-pragmatic" arguments which explicate the underlying normative presuppositions of one's own practical performance (Ott 1997).

Environmental ethics assumes that the "objective" truths of the natural sciences do not contain everything that can reasonably be said about nature. Roughly speaking, the sciences deal with nature "per se", i.e. with an objectified nature which presents itself in the same way to every neutral observer concerning its characteristics and its causal structures, while environmental ethics deals with nature "for us", i.e. with all the ways in which nature seems important, meaningful, valuable and engaging to humans. Environmental scientists should agree that nature is meaningful to many humans in extra-scientific but reasonable ways. If so, scientists within the third culture have to distinguish between two perspectives on nature. As scientists, they face nature as objectivity. As members of the third culture, they face natural sites (e.g. mires and peat lands, rivers and lakes, forests and coastlines) as units that are modified by human action in many respects and can be valued and designed in different ways. Both perspectives should not be confused but should complement each other. If so, members of the "third culture" always have to perceive nature both from an "objective" scientific perspective and from a "value laden" sustainability perspective. Such two-fold perspective does neither diminish nor distort the scientific perspective but augments and enriches it. If so, the concept of nature in environmental sciences must be a dialectical one.[2]

Environmental ethics generally asks for the reasons that should determine our individual and collective actions in dealing with non-human nature and the standards (i.e. values and norms) which are derived from these reasons. It also asks how these standards could be implemented. Therefore, environmental ethics has a theoretical and practical dimension. This chapter only deals with the theoretical dimension. In this dimension, environmental ethics asks for reasonable justifications for environmental, animal and nature conservation. Terminologically, "nature conservation" is used as an umbrella term that includes environmental conservation (e.g. water, soil, air, waste, noise, etc.), animal conservation and nature conservation *sensu strictu* (e.g. species conservation, habitat conservation, wilderness conservation, etc.). The theoretical dimension of environmental ethics is compatible with any modern conception of doing science (Ott 1997, Ch. 8). Since arguing is common to all scientific disciplines, scientists in general and members of the third culture in particular should have no principled aversion against environmental ethical reasoning.

Accordingly, the core business of environmental ethics can be understood as *critical analytics of the environmental ethical sphere of argumentation* (SA), including the presuppositions invested therein and the practical (i.e. political, legal and economic) consequences resulting thereof. Members of the third culture are invited to take an interest in this SA as such but, of

2 In this respect, it seems possible to read Hegel's often misunderstood philosophy of nature.

course, they are not committed to any substantial argument within SA. Any substantial argument is open for debate and the concept of sustainability has to be modified according to these debates. The interest in the SA results from the matter of fact that the idea of sustainability is to be warranted by SA. One may imagine the constitution of the sphere of argumentation in such a way that first of all, an empty field of possible environmental claims is opened into which reasonable answers to the basic question of *why* environmental media, living beings and certain components of nature (e.g. species, ecosystems, landscapes, etc.) *should* be conserved (i.e. protected, preserved, restored or cared for), can be entered. Intuitively acceptable answers ("conserve nature *N* because of reason *R*") are entered into the field of claims and can thus be tested against sceptical refusals. Insofar, environmental ethics takes up the intuitions, aims and corresponding speech acts of environmentalists and nature conservationists and tests them with regard to whether and, if so, how they can be transformed into sound reasons. Since members of the third culture often perform "protective" speech acts (as in the SuWaRest project: "save the Wuliangsuhai as lake", "restore the Tugai forest", "reduce pollution of the Heihe River" and the like) they are implicitly dealing with such reasons. Therefore, in the SA all known patterns of argument that "speak" in favour of nature conservation are assembled. This assembly of reasons is constitutive to environmental discourse and, as such, inescapable to any persons which participates in such discourse. Since this discourse is present within the third culture, this holds true for its members. Dialectically spoken, members of the third cultures are both invited and committed to SA. The patterns of argument that constitute the "texture" of environmental ethics can be arranged in different ways. In a classification that is oriented by the anthropocentrism vs. physiocentrism debate ("demarcation problem"), SA can be represented as shown in Table 1.

SA assembles the discourse of environmental ethics and related disciplines in a compact terminological form, which is open for any fine-grained analysis of single arguments. The bias of SA towards conservation corrects itself by presenting the reasons to sceptical persons for inspection and through the fact that all arguments can and should be critically reflected. A note on religious arguments seems appropriate at this point. Religious arguments constitute a vast array of reasoning that are based on narratives, Holy Scriptures, prayers and proverbs and spiritual practices of worshipping. Whether such "reasons" are incompatible with science, would need more elaboration than this chapter allows for. This also holds for "deep ecology" arguments. All other arguments are compatible with science since there is no scientific argument why scientists should not adopt a general attitude of reverence for life or should not feel mercy with sentient beings.

Table 1 – Sphere of argumentation of environmental ethics, adopted from Ott (2010).

A. Anthropocentric arguments
1. Dependence arguments ("livelihood", "basic needs") 2. Biophilia hypothesis 3. Health and well-being arguments ("anti-pollution" arguments) 4. Natural aesthetical arguments ("beauty", "sublime", "auratic" nature) 5. *Heimat* arguments ("Ethics of Place") ("feeling at home") 6. "Transformative value" argument ("virtuous character traits") 7. Difference argument ("civilisation should be escapable") 8. "Human-right-to-nature" argument ("right to a decent environment") 9. Obligations towards future generations with reference to 1-8. ("intergenerational responsibility with respect to all welfare effects of nature")
B. Physiocentric arguments
10. Sentientist arguments ("caring or sentient creatures") 11. Biocentric arguments („reverence for life") 12. Ecocentric arguments ("land as biotic community that includes humans") 13. Holistic arguments ("universal consideration") 14. "Deep ecology" arguments ("nature as creative and projective force", "widening identification with nature")
C. Religious arguments ("creation", "Dao", "sacred sites" and the like)

The basic question *"Why preserve nature?"*, can be answered firstly, because humans and their descendants are and (most likely) will be dependent on the ongoing utilisation of nature as resource, reservoir, sink and medium, and secondly, because certain states of nature bring about joy, pleasure, well-being, peace, delight, etc. (i.e. *Naturgenuss* as stated by Alexander von Humboldt in the 19th Century writings of *Kosmos*). Humans are not only reliant upon nature as resource but they are benefitted by the many cultural and eudemic (from "eudaimonia" which means "flourishing life" in Aristotelian ethics) values within nature. Many humans also have a morally shaped interest in the continued existence of whales, tigers, coral reefs, primary forests, etc. on this planet. Concerning such interests, economists speak of existence values. Utilisation of resources and pleasure and delight in nature can be summarised under a broad conception of human welfare. It is beyond doubt that nature contributes to human welfare in many respects.[3]

A third answer to the question for justification, starts from the intuition that nature conservation might be morally required (i.e. imposed on all persons as an understandable obligation irrespective of their cultural values, lifestyles and individual preferences) not because of its contribution to human welfare but for the sake of nature itself or for the inherent moral value of certain natural beings. Such arguments are referred to as physiocentric; arguments that address the first two answers to the initial question are referred to as anthropocentric. Therefore, conceptions of environmental ethics deal with nature as a resource, nature as a source of pleasure and nature as an ensemble of beings that might have inherent moral value. Physiocentric arguments agree that natural beings have inherent moral value, but disagree on the scope of such entities. After decades of debate, the mainstream-solution is sentientism (Krebs 1999; Ott 2008). To sentientism, all sentient beings count morally as such. Whether moral respect to sentient beings might (not) be graded and might be applied to domesticated and wild-living sentient beings, is beyond the scope of this chapter. In any case,

3 The ecosystem services approach is a way of how to order welfare effects. Studies that exploit the ecosystem services approach have come to the conclusion that cultural values of nature contribute far more to social welfare than conventional economic wisdom has made us believe (Jax et al. 2013).

adoption of sentientism implies to include animal welfare in the concept of sustainability. Thus, birdlife at Wuliangsuhai would count not only with respect to the delight bird watching brings about. Delight in bird watching and the inherent moral value of birds constitute both the moral significance of birdlife.

SA as such contains neither criteria for the solution of conflicts in nature conservation nor a casuistry for evaluating special cases in detail (as in Wuliangsuhai and the Heihe River Basin). Both are topics of single projects within environmental disciplines, which touch real-world conditions at specific study sites. Furthermore, it does not contain a certain conception of what it might mean to "weigh" issues of nature conservation with other issues, as economic and societal ones (e.g. urbanisation, meat production, energy consumption and tourism). SA rather makes all persons, including policy-makers and members of the third culture more aware about the many conflicts that occur if reasoning about nature's values is taken seriously. In the first instance, environmental ethics multiplies conflicts. The questions of "good" conflict solving, "careful" consideration and "appropriate" assessment of individual cases presuppose a thorough examination of the SA because otherwise, the conventional thought patterns remain dominant and concerns of nature conservation may be "weighed away", as it is all too often the case. Members of the third culture often are in the middle of such conflicts. The ethos of the third culture might require them to take the role of environmentally concerned scientists and, as such, to advocate long-term human and even nature's interests and to look for feasible and viable strategies for adaptive management and for a transformation towards sustainability. Such advocacy is always in tension with principles of presumptive neutrality and with the commitment not to interfere with governmental affairs of foreign countries in which research often will be been performed (as in SuWaRest). In domestic affairs, members of third cultures can provide recommendations of how to act to policy-makers; in foreign countries, they may restrict themselves to provide scenarios (see Chapter 10). Thus, advocacy for sustainability must be tempered by prudence and politeness.

1.3 The concept of sustainability

So far, members of the third culture are committed only to the very idea of sustainability but not to any specific concept. From sustainability discourse, however, they may recognise that there are competing concepts of sustainability. Members of the third culture are also committed to the overall discourse in environmental ethics but not to any single substantial argument. They might recognise some broad and general insights that have resulted from some decades of debate, as (1) the many welfare effects of nature and (2) the mainstream solution with respect to the demarcation problem (i.e. sentientism) but they remain free to challenge and question any argument. If members of the third culture agree up to this point, they should take an interest in both SA and the contest between different concepts of sustainability. We turn now to such concepts of sustainability.

In its core, the idea of sustainability has been conceived rather anthropocentrically. Nature is taken fully into account as a source of welfare, joy and meaning and it is assumed that members of future generations will, with high likeliness, also be benefitted by the many values within nature. All items and sites of nature, which bring about welfare effects or so-called ecosystem services fall under the term "natural capital". To the SuWaRest project, Heihe River, Tugai forests and Wuliangsuhai Lake they have been perceived as critical natural capital of Inner Mongolia. The critical question against such value-based perception is grounded in the possibility to replace and substitute natural capital by artificial capital, by technologies, and by commodities that benefit humans. Since many economists are familiar with substitution processes in both production and consumption they might not deny welfare effects of natural capitals but might cast doubts on claims that these welfare effects outweigh other kinds of welfare effects, as monetary income. Since environmental economics is part of the third culture, casting such doubts is perfectly legitimate and must be addressed. SuWaRest, for instance, faced the problem whether agriculture production at Heihe middle-stream and in Hetao Irrigation District outweighs the benefits of Tugai forest and Wuliangsuhai Lake. Clearly, humans can conform and cope with many artificial environments and one cannot predict with certainty how deeply different cultures may feel the loss of natural goods, as, for instance, with the Tugai forests in the Ejina region or Wuliangsuhai Lake within the League of the Bayannur. Perhaps, most Chinese people feel comfortable with cheap food, skyscrapers, shopping malls and other items of urban life. If so, substitution of nature is always an option to any advanced society. One scenario, as being presented in Chapter 10, substitutes the Wuliangsuhai Lake by a wetland that is designed for "sustainable" reed production. In principle, the psychological welfare effects of a water-consuming space-flight station as being located at midstream of the Heihe River Basin to many Chinese people may outweigh the existence value of Tugai forests at its downstream. Therefore, serious economic topics like substitutability of welfare effects show up within the third culture. Scholarly persons can elucidate on them but, ultimately, it must be decided by politic affairs. Members of the third culture can point at such loss of nature but must leave the decision to stakeholders and policy-makers, hoping for comprehensive environmental deliberation on such matters.

Generally, citizens of different societies must specify the basic question "*What to sustain?*" with respect to different types of capital (i.e. human made capital, human capital and natural capital). They should not leave the answer to market forces since real markets function in ways that privilege commodities over collective goods. Any answer on this basic question that refers to a fair bequest package will rest on assumptions that are contested within the ongoing theoretical debate on "weak", "intermediate" and "strong" sustainability, such as substitution, technological progress, discounting and compensation. Very weak sustainability is about growth of gross domestic product, weak sustainability permits substitution of natural capital if the overall genuine savings of a society are positive, intermediate sustainability requires to preserve all natural capital which is "critical" in some respect to be defined, strong sustainability is committed to hold natural capital at least constant and, finally, very strong sustainability adds animal welfare to the picture because sentientism is adopted from SA. For the sake of my overall argument, I do not wish to commit all members of the "third culture" to a specific solution of the inherent-moral-value-problem (i.e. a "demarcation problem"). It might be sufficient to make scientists aware of the very problem that plays a role in any Mode-II project in which wildlife and animal farming plays a role.

This ongoing debate on concepts of sustainability constitutes a vast array of claims, arguments, refutations, scientific evidence, models and the like. Given discursive freedom, different societies may reach different conclusions about the amount of natural capital that should be preserved, about criticality of natural capital, about the contribution of ecological services to societal welfare, about the rate of discount and about the degree of substitutability between different types of capital. In the first instance, any decent society is entitled to adopt freely whatever concept of sustainability they believe to be "superior" or "more favourable". If the discursive procedure has been fair, the outcome has to be respected fully on political grounds. This condition of fair discourse is, of course, not fulfilled in many countries. Very often, substitution is not chosen deliberately by people but continues as a brute economic force that is imposed upon them. Therefore, it may happen that "third-culture"-scholars reject substitution of natural capital but perform their research in a country within which natural capital is substituted by human-made capital at high speed. In my opinion, this was the case in the SuWaRest project.

Matters would look highly different, however, if scholars and people would reach some common moral ground and would agree that the theoretical debate between weak, intermediate and strong concepts of sustainability has, indeed, provided some robust results in favour of, at least, strong sustainability. Such results emerge if questions are framed with respect to SA in general, not only with respect to economic utilisation of nature as a resource. The crucial question is not only "*Can we substitute nature by technical means?*" nor "*Can we substitute natural resource inputs for production?*", but also "*Do we really wish to substitute natural environments by more artificial ones if nature has many non-material, cultural welfare effects on many of us?*" Human life will continue if natural capital will be depleted but it might not be desirable to do so. With respect to nature conservation, single groups within societies may argue from within strong environmental traditions (as Daoism in China) or even with moral convictions about inherent moral values in nature. The question to supporters of very strong sustainability then is "*Are we morally permitted to substitute natural sites that serve as habitat for sentient wildlife?*"

There are arguments about risk and precaution with respect to critical natural capital. If intermediate sustainability requires maintaining all critical natural capital and if there are large uncertainties about criticality, one should better adopt an ambitious safe minimum standard. If prudent members of decent societies ask themselves "*How safe is safe enough?*" and if they look back to a long historical process of domination, exploitation and over-utilisation of nature they better should conclude that they should maintain *prima facie* the remaining stocks of natural capital. If so, intermediate sustainability tends towards a Constant Natural Capital Rule (CNCR), which is constitutive to strong sustainability.

There is a cultural dimension of uncertainty as well. Uncertainty of future preferences, if taken seriously, should make any society more cautious against depletion of nature since many members' future generations may be more sensitive to nature's values and might be more open even to spiritual encounters with nature as ordinary "rationalised" members of Western civilisation have been throughout the 20th Century. We should not rule out the possibility that future people may have rediscovered the evolutionary biophilic disposition of humans and may shape this disposition in new cultural ways, wishing to live lightly in nature. The human capability of being able to live with a deep concern for a world of nature, including animals and plants, might be actualised to high levels by future individuals. If one hopefully assumes that the habits and attitudes of future generations might be shaped by SA, it would be absurd to bequeath a highly artificial world to them as result of ongoing substitution processes.

1.4 Conclusion

Therefore, prudent and long-term oriented anthropocentrism can make a strong case in favour of concepts of sustainability, which demand to hold natural capitals constant over time for the sake of future generations with respect to the many valuable benefits of nature (Daly 1996; Ott & Döring 2008; Ott 2014). In other words, different arguments motivate a reasonable and prudent choice at least in favour of strong sustainability. Such choice is clearly not a proof but rather a judgment. It is not strictly binding, as proofs are, but favourable to members of the third culture and recommendable to any society. As judgment, it constitutes a collective *prima facie* obligation to hold natural capitals (i.e. natural goods) at least constant over time (i.e. CNCR). CNCR, being the basic rule of strong sustainability, gives content to the problem of how to conceive a fair intertemporal bequest package. CNCR must be specified to a comprehensive system of rules (so-called "management rules"). The establishment of a rule-based governance scheme, which specifies the CNCR, defines the core meaning of institutionalising strong sustainability. Holding natural capital constant over time should reduce pollution, while the undertaking to restore a depleted reserve becomes mandatory. The SuWaRest project took such a perspective within its study areas, which can be generalised as third culture.

If the argument were sound, the concept of strong sustainability would ground the interests of knowledge within environmental sciences and within the third culture. Such grounding gives a focal epistemic perspective to the third culture and it explains the role of "concerned scientists".[4] To many scientists, these arguments looks as a strong and uncommon claim. Scientists within the third culture may feel uncomfortable with this claim since it seems to be a new way of "moralising" science. They may fear to be pressed *nolens volens* into political alliances with environmentalists and conserva-

4 In 2010, at a meeting in Zhangye, the SuWaRest project confronted a somewhat complicated task of translating "concerned scientists' viewpoints" to our Chinese counter partners. Such translation was not only a linguistic issue, but also touched deeper problems about the role of environmental science in China, which is often solely reduced to data mining.

tionists. Therefore, anyone who is engaged in environmental sciences should feel challenged by this claim in order to refute such transcendental-pragmatic grounding.

Key references

Habermas, J. 1965. Erkenntnis und Interesse. In: A. Hans & E. Topitsch eds. Werturteilsstreit. Darmstadt: Wissenschaftliche Buch Gesellschaft, 1990, 334–352.

Krebs, A. 1999. Ethics of Nature. Berlin: DeGruyter Verlag.

Ott, K. & Döring, R. 2008. Theorie und Praxis starker Nachhaltigkeit. Marburg: Metropolis-Verlag.

Ott, K. 2008. A Modest Proposal of How to Proceed in Order to Solve the Problem of Inherent Moral Value in Nature. In: L. Westra, K. Bosselmann, & R. Westra eds. Reconciling Human Existence with Ecological Integrity. London: Earthscan, 39–60.

Ott, K. 2014. Institutionalizing Strong Sustainability: A Rawlsian Perspective. Sustainability, 6(2): 894–912.

Ott, K. 1997. Ipso Facto. Frankfurt: Suhrkamp Verlag.

Ott, K. 2010. Umweltethik zur Einführung. Hamburg: Junius Verlag.

Snow, C.P. 1959. The Two Cultures. Canto 1993. London: Cambridge University Press.

Snow, C.P. 1990. The Two Cultures. Leonardo, 23(2): 169–173.

2. Study areas: The Heihe River Basin and Wuliangsuhai Lake at the Hetao Irrigation District

Niels Thevs
Konrad Ott
Lilin Kerschbaumer
Ping He

2.1 Introduction

Central Asia, which extends from the Caspian Sea to northwest China and Mongolia, is largely covered by deserts, steppes and mountain ranges. Within this huge dryland area, there are numerous wetlands, either distributed along river systems or island-like dispersed in depressions. In terms of area, the former predominate over the latter. Naturally, there is a mosaic of mainly *Phragmites australis* dominated wetlands and riparian forests distributed along the rivers of Central Asia (Ogar 2003). These wetlands and forests play a crucial role for the biodiversity of that region, because they offer habitats for wildlife and most plant species in the drylands of Central Asia (WLI 2012). Ever since, the rivers have attracted people to settle and establish themselves around oases, as this warranted a water supply in the dryland areas of Central Asia. Some of those oases have a history of several thousands of years and are part of the Silk Road, like Buchara and Samarkand, Kashgar, Hotan or Zhangye; however, during the past six decades, the speed of land reclamation and expansion of settlements increased tremendously. Huge areas of the natural ecosystems along the rivers have been converted into either agricultural land or settlements or have been degraded due to water shortage. The most prominent example is the expansion of cotton in the former Soviet Union, which resulted in the desiccation of the Aral Sea and degradation of wetlands along the Amu Darya River (Glantz 2005). A similar example in China is the Heihe River.

The downstream section fell dry in the 1970s in the Hetao Irrigation District, a riparian wetland complex along the Yellow River has been converted into agricultural land. *P. australis* played and still plays an important role for people in Central Asia. It used to be, and in parts of the region still is, the major fodder plant for livestock (Thevs et al. 2007).

Traditionally, it was used for mats in house construction. Today, it plays a role as raw material for e.g. paper production or insulation material (Köbbing et al. 2014a), and as energy source (Patuzzi et al. 2013a). Furthermore, as population along the rivers of Central Asia increase and settlements expand, wetlands play an increasingly important role with respect to water purification. Against this backdrop of degradation of wetlands in Central Asia, from past to present, their significance throughout the whole region of Central Asia, a wide range of initiatives has been undertaken in order to protect and restore wetlands. Examples are protected areas, which have been established during the past 30 years, e.g. Nizhny Amu Darya Biosphere Reserve, Uzbekistan; Amu Darya State Reserve, Turkmenistan; Ili Delta Nature Reserve, Kazakhstan; Tarim Huyanglin National Nature Reserve, Xinjiang, China; and Ejina Huyanglin Nature Reserve, Inner Mongolia, China. Wetland protection in a dryland region, like Central Asia, is closely connected with the water resource management of the rivers, which sustains particular wetlands.

The vast majority of the rivers in Central Asia face upstream-downstream conflicts over water. Upstream countries, regions or water users divert and consume water at the cost of downstream riparian countries or regions. When the cotton production was promoted in the previous Soviet Union, it occurred at the cost of the Aral Sea as well as the Amu Darya and Syr Darya deltas. In a similar way, at the Tarim River water users upstream consume water with the result that downstream users suffer water shortages. The Heihe River Basin represents such upstream-downstream conflicts over water distribution in river systems. Wuliangsuhai Lake is a showcase for a wetland suffering eutrophication in a context of an irrigation scheme – via the Hetao Irrigation District. Hence, two key points of examination arise: (1)

in the former, water quantity and allocation of certain water amounts and (2) in the latter, just water quality. Therefore, both sites are showcases for the water allocation and water quality problems, which are prevalent throughout Central Asia.

2.2 The Heihe River Basin

The Heihe River Basin covers an area of 120,000 km^2 and is shared by Gansu Province and Inner Mongolia. The headwaters of the Heihe River are located in the Qilian Mountains in Gansu Province south of the city of Zhangye (Figure 1). The Heihe ends in the two terminal lakes West and East Juyanhai Lake close to the border with Mongolia (Li et al. 2012a). From the Qilian Mountains, the Heihe River flows into the oasis of Zhangye, which has a population of about 1.3 million inhabitants. Within the area of Zhangye, about thirty small rivers flow down from the Qilian Mountains, which now are diverted into an irrigation zone. Only during spring or high floods some of these rivers contribute to the Heihe River's runoff. Once the Heihe leaves Zhangye, it flows as a so-called losing stream through mainly gravel deserts northwards from Gansu into Inner Mongolia. This is a common feature of rivers in the whole of Central Asia, in which they originate from mountain areas due to a surplus of precipitation, i.e. rain and snow. The rivers thus are fed by rainfall and melt-water from snow and glaciers alike. Thereby, the runoff of the rivers on the western side of the Tianshan and Pamir mountains is melt-water dominated, e.g. Amu Darya and Syr Darya. In contrast, rainfall contributes to a significant part of the runoff of rivers east of the Tianshan and Pamir, like the Tarim and Heihe. In the Tarim and Heihe river basins, the precipitation maximum is in summer, while the former river basins receive most precipitation in autumn and spring. In higher elevations, where the rivers originate, spring and autumn precipitation falls as snow.

Once rivers like the Heihe flow away from the mountains where they originate, the rivers turn into so-called losing streams. This means that such rivers constantly lose water into adjacent groundwater aquifers or via evaporation (Hou et al. 2007). Parts of rivers throughout Central Asia vanish in the desert due to this water loss, like the Keriya or Niya in the Tarim Basin or the Chu River in Kazakhstan. Other rivers, like the Amu Darya, Syr Darya, Ili, Tarim or Heihe become smaller, further downstream, and drained or still drain into terminal lakes. So all rivers in Central Asia, except the Irtysh, are endorheic rivers, which means that they do not reach the ocean. Thus, the Heihe River represents an endorheic river basin of Central Asia.

The climate in the Heihe River Basin is arid and continental. In the Qilian Mountains along the headwaters of the Heihe River, the annual precipitation is about 400 mm. In Zhangye at the foothills of the Qilian Mountains, it is 170 mm and further north in Ejina it is only 60 mm. About two thirds of the annual precipitation is concentrated in the months from June to August. This precipitation is maximised when it falls together during the snow and glacier-melting period in the Qilian Mountains, which results in annual summer floods in the Heihe River. Such summer floods naturally occur in all rivers of Central Asia.

The natural vegetation along the Heihe River consists of a mosaic of riparian forests and reed beds dominated by *P. australis*. The largest reed beds are located around the two terminal lakes of the Heihe. This area was an important pasture ground for Mongolian herders. The Mongolian people here are a minority within China. Today, reed beds can be found around the eastern terminal lake and smaller patches along the Heihe. In Zhangye, a wetland park has been established, in order to create a recreation site. Furthermore, small reed stands are distributed all over Zhangye, which receive wastewater and play an important role in purifying wastewater.

Due to the arid climate, all agriculture along the Heihe River depends on irrigation. The history of irrigation in Zhangye has been documented for more than 2000 years (Feng & Cheng 1998). During that time, no irrigation agriculture was known along the lower reaches of the Heihe River in present day Inner Mongolia. Starting in the 1950s, the area under agriculture along the Heihe, in Zhangye, was enlarged like in all other oases in northwest China (Gruschke 1991) from 82,600 ha in 1949 to 260,000 ha in 1995 (Feng & Cheng 1998) and 253,300 ha in 2012 (personal communication with the agriculture administration of the city of Zhangye 2012). Cropland also was reclaimed in Ejina County. While before 2000, the major crops along the Heihe were cotton and paddy rice, now the major crop is seed corn. Most of the seeds, which are used to crop corn in China, are produced in Zhangye.

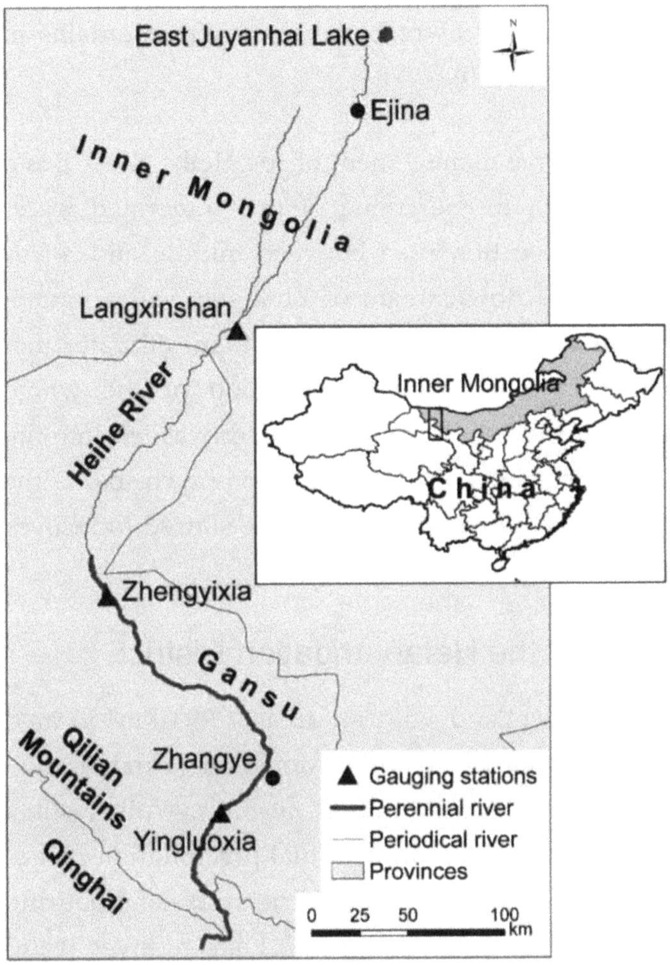

Figure 1 – Map of the Heihe River Basin, China (Liu 1997).

This increase of cropland area resulted in increasing demand for irrigation water, which was diverted from the Heihe River. The terminal lake West Juyanhai has been dried out since 1961 as well as most of the western branch of the Heihe in Ejina. The terminal lake East Juyanhai Lake covered 35.5 km^2 in 1958, shrunk to 23.6 km^2 in 1980 and dried up completely in the beginning of the 1990s (Ejina Qizhe 1998). In 2002, it reappeared for a few months with an area of 12 km^2 (Wang et al. 2002). In the course of decreasing runoff reaching Ejina County, soil salinisation increased in parts of the county (Qi & Cai 2007). Groundwater levels dropped from 0.5 – 1.3 m in the 1940s to 3 – 6 m in the 1990s (Guo et al. 2009). However, the runoff from the Qilian Mountains has not changed significantly during the past 50 years despite climatic changes and shrinking glaciers as reported for other mountains in China and Central Asia (Jiang & Liu 2010).

In 2000, an integrated water resource management of the Heihe River Basin was established (Guo et al. 2009). In the frame of this integrated water resource management a water allocation plan between middle and lower reaches, i.e. between Zhangye and downstream of Zhangye, was adopted, which is described in detail in Chapter 4. Today, the amount of water that Ejina receives lies above the annual evapotranspiration of the whole cropland and riparian ecosystems within Ejina. More water was led into the eastern river branch of the Heihe compared to the western branch so that the reed and shrub vegetation around the East Juyanhai Lake started to recover (Guo et al. 2009).

2.3 Wuliangsuhai Lake and the Hetao Irrigation District

Wuliangsuhai Lake is a shallow wetland with an area of 293 km^2 located about 100 km west of the city of Baotou in Inner Mongolia (Yamian et al. 2012). More than half of the wetland is covered by *P. australis*. Wuliangsuhai Lake is a wetland in a dryland region. The mean annual precipitation ranges from 139 mm to 222 mm in the Hetao Irrigation District and neighbouring Wuliangsuhai Lake (Wang et al. 2004). Wuliangsuhai Lake receives water from the neighbouring Hetao Irrigation District. Thus, it is a representative example for the many wetlands in Central Asia, which are located, downstream of an oasis and are sustained by the drainage waters of those oases. Looking at the map of China, the Yellow River forms a great loop in the three provinces Ningxia, Inner Mongolia and Shanxi. The Yellow River flows in a northeast direction from Ningxia into Inner Mongolia. There, the Yellow River turns east for about 350 km. East of Baotou, the river turns south and flows from Inner Mongolia into Shanxi Province. In the curved area, where the Yellow River turns east in Inner Mongolia, a major river course, the Wujia River branches off from the Yellow River and goes back into the Yellow River, precisely, Wuliangsuhai Lake. Several river courses connect the Yellow River's mainstream and the Wujia River so that an inland delta with numerous wetlands are formed. At present, Wuliangsuhai Lake is the most eastern part of this inland delta (Wu et al. 2013; Fejes et al. 2008).

As all other rivers with their accompanying wetlands in Central Asia, this inland delta has been attracting people ever since. Agriculture is documented from the Han Dynasty (206 BC to 220 AD), and according to the strength of the ruling Chinese dynasties, it expanded during periods of strong leadership (e.g. Tang and Song Dynasties) and shrunk during periods of weak central power. This inland delta has also been an important region for Mongolian nomadic herders. During the end of the 19th Century, under the Qing Dynasty, eight main irrigation channels were constructed, which laid the basis for the irrigation scheme of today in the Hetao Irrigation District. In 1923, the railroad from Beijing via Inner Mongolia to Lanzhou was constructed leading to an increased migration into Inner Mongolia and further land reclamation in the Hetao Irrigation District (Wu et al. 2013; Fejes et al. 2008). In 2005, the Hetao Irrigation District contained 5,860 km^2 of farmland, including a number of garden plots, which increased to 5,900 km^2 in 2010 (BLRB 2012). The water for the irrigation of the whole district is diverted from the Yellow River at the Sanshenggong Water Station in the south western tip of the Hetao Irrigation District into the central irrigation channel, which runs parallel to the Yellow River. From that channel, the water is further diverted through a network of 7,645 km of irrigation channels arranged around ten main channels. Parallel to the structure of the main irrigation channels, there is a network of 2,535 km of drainage channels. The central drainage water collector runs in the previous river course of the Wujia River (Figure 2). This central drainage collector enters into the northern part of Wuliangsuhai Lake and is the main water source for the lake. From 1960 to 1980, annually 4–5 km^3 water were diverted from the Yellow River into the Hetao Irrigation District (Yamian et al. 2012). This amount of water increased to 5-6 km^3 after 1980 until today. From the Hetao Irrigation District, 0.1–1 km^3 water was drained into Wuliangsuhai Lake from 1960 to 1980. After 1980, this amount of water increased to 0.5–1.2 km^3 per year. The annual water intake of the Hetao Irrigation District and the runoff into Wuliangsuhai Lake are further explored in Chapter 10. The main crops planted in the Hetao Irrigation District are rape, wheat, corn and sunflower on 1,720 km^2, 1,630 km^2, 1,510 km^2 and 1,290 km^2, respectively. Further crops are sugar beet, melons and tomatoes.

Wuliangsuhai Lake, unlike the terminal lakes of the Heihe, has never fallen dry. It continuously receives water. As the water source for Wuliangsuhai Lake is drainage water from the Hetao Irrigation District, it receives an annual load of 2,292.65 t of total nitrogen and 247.36 t of total phosphorus from agriculture, industry and households in the Hetao Irrigation District (BCPG 2010). Therefore, Wuliangsuhai Lake suffers from eutrophication. It is a showcase for a wetland downstream of an intensively agricultural cropped oasis. The core issue of Wuliangsuhai Lake, therefore, is water quality rather than water quantity.

Wuliangsuhai Lake formed in a shallow depression as part of the previous inland delta of the Yellow River. Today, its area is between 293 km^2 and 310 km^2, depending on the water level. The average water depth is 1 m with a maximum water depth of 4 m. The water volume ranges between 0.25 and 0.3 km^3 (Yamian et al. 2012; Liu et al. 2007b; Fejes et al. 2008). The climate is arid and extremely continental with an annual precipitation of 222 mm, a minimum temperature of -38 °C in January, and a maximum temperature of 38 °C in July. The mean annual temperature is 7.3 °C. Due to the low winter temperatures, the lake is frozen from November to April for 152 days on average (Yamian et al. 2012; Faafeng et al. 2008; Fejes et al. 2008). Due to the nitrogen and phosphorus input brought into Wuliangsuhai Lake by the drainage waters from the Hetao Irrigation District, the lake suffers from severe eutrophication and, accordingly, deterioration of water quality.

The dominant vegetation in Wuliangsuhai Lake are reed beds from mainly *P. australis* and to a limited extent from *Typha latifolia* (Zeng et al. 2012). Those reed beds cover 188 km^2, more than half, of the lake (Shang et al. 2011). Enhanced by eutrophication, the reed bed area increased from 165 km^2

in 1986 to 188 km² to date. Along with the decreasing water quality, the amount and area of *Potamogeton*-dominated submerged vegetation and algae increased. The biomass from submerged vegetation and algae deposits on the sea floor every autumn, which gradually reduces the water volume of the lake. In addition, biomass from *P. australis* and *T. latifolia*, if not harvested, tend to build up, creating deposits and contributive masses that gradually reduce the lake's overall water volume. Recently, the annual rate of deposition is 2 cm (Zeng et al. 2012).

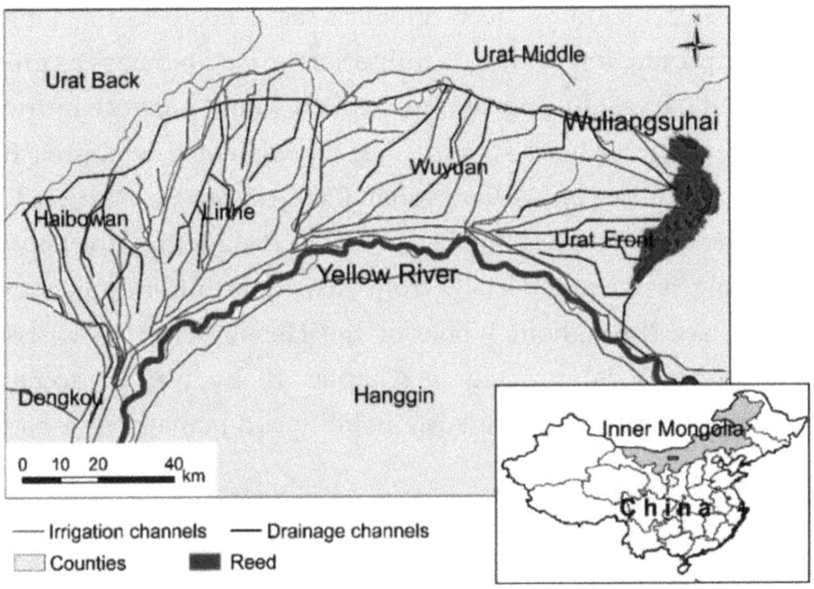

Figure 2 – Map of the Hetao Irrigation District and location of Wuliangsuhai Lake (Liu 1997).

Wuliangsuhai Lake serves as water storage reservoir for the Yellow River by discharging water into the Yellow River during the low water period in spring. Therefore, the water volume available for water storage is of importance for the whole Yellow River basin downstream of Wuliangsuhai Lake. In addition, the quality of the water released from Wuliangsuhai Lake into the Yellow River is of importance for water users downstream (Wu et al. 2013). Furthermore, the lake and its reed beds provide habitat for migratory and breeding birds (Faafeng et al. 2008). In 1993, Wuliangsuhai Lake became a provincial nature reserve (Zeng et al. 2012).

In Wuliangsuhai Lake, reed plays a crucial role for regulating the water quality. *P. australis* has a high ability to purify water as investigated in Chapter 6. Harvesting reed, *Potamogeton*, and algae biomass also may remove nutrients from Wuliangsuhai Lake (Frick et al. 2011). *P. australis* and *T. latifolia* are harvested in winter, when harvesters can access the reed easily on the frozen lake. This biomass is sold as raw material for paper production as described and analysed in Chapter 8. In addition, more options for reed biomass utilisation are explored in Chapter 8 (Köbbing et al. 2013). The option to use reed biomass as energy source is analysed in Chapter 7 and Patuzzi et al. (2013), including reed biomass as feedstock for biogas production. Biogas production as well as utilisation of reed biomass as green manure require reed harvest during summer, which would enhance nutrient removal compared to winter harvest, but is less convenient. The approaches to tackle the eutrophication of Wuliangsuhai Lake cannot be restricted to measures within the lake. There is a holistic need to include the diversity of agricultural cropping systems, especially from varying fertiliser applications currently put in place throughout whole of the Hetao Irrigation District. Such a holistic approach is explored in Chapter 10 by way of scenario alternatives, opening up pathways for sustainability and management-based objectives.

Key references

Ejina Qizhe. 1998. Ejina Qizhe (Description of Ejina County). Beijing: Fangzhe Chubanshe.

Faafeng, B., Li, T., Lindblom, E., Ye, J., Oredalen, T.J., Lövik, J.E.L. & Svenson, A. 2008. Lake Wuliangsuhai Restoration Project: Water Quality Monitoring System. Norwegian Agency for Development Cooperation Agency.

Fejes, J., Ratnaweera, H., Yawei, L., Lindblim, E. & Faafeng, B. 2008. Inner Mongolia Lake Restoration Project, Lake Wuliangsuhai Comprehensive Study Extension, Final Report. Norwegian Institute for Water Research.

Feng, Q. & Cheng, G.D. 1998. Current situation, problem and rational utilisation of water resources in Gansu Province. Chinese J. Arid Land Research, 11: 293–299.

Guo, Q., Feng, Q. & Li, J. 2009. Environmental changes after ecological water conveyance in the lower reaches of Heihe River, northwest China. Environmental Geology, 58(7): 1387–1396.

Köbbing, J.F., Thevs, N. & Zerbe. 2013. The utilisation of Reed (*Phragmites australis*) – A review. *Mires and Peat*, 13: 1–14.

Thevs, N., Zerbe, S., Gahlert, E., Mijit, M. & Succow, M. 2007. Productivity of reed (*Phragmites australis* Trin. ex Steud.) in continental-arid NW China in relation to soil, groundwater, and land-use. *Journal of Applied Botany and Food Quality-Angewandte Botanik*, 81(1): 62–68.

Zeng, Q., Zhang, Y., Jia, Y., Jiao, S., Feng, D., Bridgewater, P. & Lei, G. 2012. Zoning for management in wetland nature reserves: A case study using Wuliangsuhai Nature Reserve, China. *SpringerPlus*, 1(1): 23.

3. Use of microbial communities as bioindicators for land-use systems

Lorenzo Brusetti
Luigimaria Borruso

3.1 Introduction

The release of industrial and municipal waste products in freshwater ecosystems has become a dramatic issue for the environment and for human health. Freshwater contaminants are mostly associated with the particles transported by water, attached to their surfaces through covalent as well as ionic chemical bonds. In slow-flowing water, particles can be deposited by gravity into sediments, where particular biological and chemical properties lead to microbial communities to form a complex biofilm in oxic as well as anoxic conditions. Thus, freshwater sediments can host Bacteria, Archaea and other microorganisms capable of a plethora of different metabolisms, ranging from anaerobic respiration to fermentation, dehalogenation, oxygenic and anoxygenic photosynthesis and others.

These microbial communities react to any compounds released into the freshwater system by increasing or decreasing of taxa, and by genetic richness and diversity. Microbial communities are, indeed, able to provide an effective and integrated measure regarding the presence and effects of toxic xenobiotics in water.

3.2 Biomonitoring in a freshwater environment

The standard chemical-physical analysis of freshwater environments has been historically based on the monitoring of parameters such as the Biochemical Oxygen Demand (BOD), Chemical Oxygen Demand (COD), measure of Total Suspended Solids and determination of the concentration of metals, nutrients, organic compounds or xenobiotics. These parameters provide information about the concentrations of a particular contaminant in an ecosystem, but they do not inform on the additive, antagonistic and synergistic effects of such compounds and elements on the biota (McGeoch 1998; Fränzle 2006). Moreover, an analytical approach gives insight on the sampling time, and does not permit detection of sporadic release or presence of pollutants. Analytical investigations cannot reflect the integration of numerous environmental variables because freshwater environments are usually strikingly complex, in which multi-sources of numerous contaminants are often rapid and difficult to estimate any hydrological changes. In this context, a study of bioindicators can be the best choice to assess ecosystem quality.

Bioindicators are biological processes, species or groups of species used to monitor biotic and abiotic variations of an environment during a certain span of time. Bioindicators show the cumulative impact of pollutants on biota, providing relevant long-term information on the environmental status or trends (McGeoch 1998). An ideal bioindicator should show a measurable and proportional response to environmental stresses. It has to be widely spread in the study area and stable, despite moderate environmental variations. It should have a low mobility, to avoid its movement in long distances, far from the pollution source. Finally, it has to be easily sampled and classified to avoid excessive time, i.e. the use of highly specialised personnel and prohibitive analytical costs.

Basically, there are two different approaches in biomonitoring. In the first case, organisms, which already exist in the environment, are observed and analysed to provide information about the environmental status (Boothroyd & Stark 2000; Parr & Mason 2003). In the second case, bioassay organisms either are used in the laboratory to test for an example of toxicity via an environmental sample, or directly introduced on-site to monitor the overall environmental quality (Girotti et al. 2008). For example, periphyton, a complex matrix of algae, fungi, protozoa, metazoa and heterotrophic microbes diffused in almost all aquatic ecosystems, can be considered as pointer multi-assemblage of organisms. It is commonly attached on rocks and on other submerged substrates, playing an important ecological role in freshwater food webs (Rosemond et al. 1993). Periphyton can accumulate many types of pollutants, ranging from heavy metals to bacterial pathogens. These pollutants can remain stable and protected for a long time in the biological matrix (Ács et al. 2003). Algae respond to an environmental stress, such as nitrogen and phosphorous surplus, by decreasing their diversity and richness, or changing their taxonomic composition, or varying their biomass. Other organisms, such as aquatic invertebrates, are commonly used to assess freshwater quality because of their long-term presence in sediments and sensitivity to changes in water or habitat quality. The presence and absence of taxa or variations in their richness and diversity are related to the occurrence of organic pollution levels (Boothroyd & Stark 2000; Parr & Mason 2003).

As mentioned above, bioassays may be an alternative approach in biomonitoring. For instance, measure of the mortality level of *Daphnia magna*, a freshwater zooplankton species belonging to the subphylum Crustacea, is widely used to determine the pollution level of freshwater samples, since *D. magna* is easier affected by ingestion of xenobiotics than other organisms. Moreover, it is a good candidate for bioassay analyses to monitor residuary waters, due to its short generation time, high rate in multiplication, and easiness of manipulation and maintenance in laboratory conditions. Furthermore, its physiological answer to toxicity can be evaluated in a relatively short time (Villegas-Navarro et al. 1999; Emmanuel et al. 2004).

Microorganisms are considered good bioindicators because they respond quickly to physical and chemical environmental changes. Bacteria, fungi and Archaea are strictly interconnected with the surrounding environment, because of their high surface area-to-volume ratio. Bacteria are more abundant in a microsite compared to other microorganisms and their assay is becoming ever more rapid and economically advantageous. Their analysis is particularly advantageous since a huge amount of individuals can be harvested and processed from a very small volume of sample, minimising the sampling disturbance.

Bacteria can be used as bioindicators at different levels: gene, population and communities. They have evolved resistance mechanisms against contaminants, which involve specialised functional genes. The abundance and occurrence of the above-mentioned genes can provide evidence of the presence of certain xenobiotics such as antibiotics, heavy metals and polycyclic aromatic compounds (PAH), which might be difficult to identify through routine measurements. In agricultural areas characterised by copper pollution, it was observed a widespread presence of bacterial strains characterised by the occurrence of the *cop*-gene family, which confers resistance to copper. On the contrary, the authors did not detect *cop*-genes in the samples collected in non-polluted sites (Altimira et al. 2012). Fuel combustion, waste incineration, coal gasification and petroleum refining processes produce a large amount of PAH that can be detected through the analysis of PAH-specific ring hydrolysing dioxygenases (RHD), in which the involved genes are a part of the cleavage of the ring in an aromatic hydrocarbon (Figure 1a) (Kumar & Khanna 2010).

A number of studies have reported that there is a correlation between antibiotic-resistance genes and the levels of antibiotic concentrations in the environment (Wu et al. 2010; Allen et al. 2010). Wu et al. (2010) found a link between tetracycline concentration and the presence of tetracycline resistance (*tet*) genes in the vicinity of nine swine farms located in three cities in China. Potentially heavy metals and others toxic compounds such as quaternary ammonium compounds, antifouling agents and detergents may select for genes encoding antibiotic-resistance (Figure 1a) (Berg et al. 2005; Singer et al. 2006). Genes encoding heavy metals resistance can be located together with antibiotic-resistance genes. Bacteria may have also an unspecific mechanism of resistance common to different substances, for example, the multivalent pumping systems prevent the intracellular accumulation of structurally diverse xenobiotics (Figure 1b) (Piddock 2006; Poole 2005). A high concentration of antibiotic-resistant genes were detected in agricultural soil treated with copper and in freshwater microcosms with a high concentration of heavy metals (Berg et al. 2005; Stepanauskas et al. 2006).

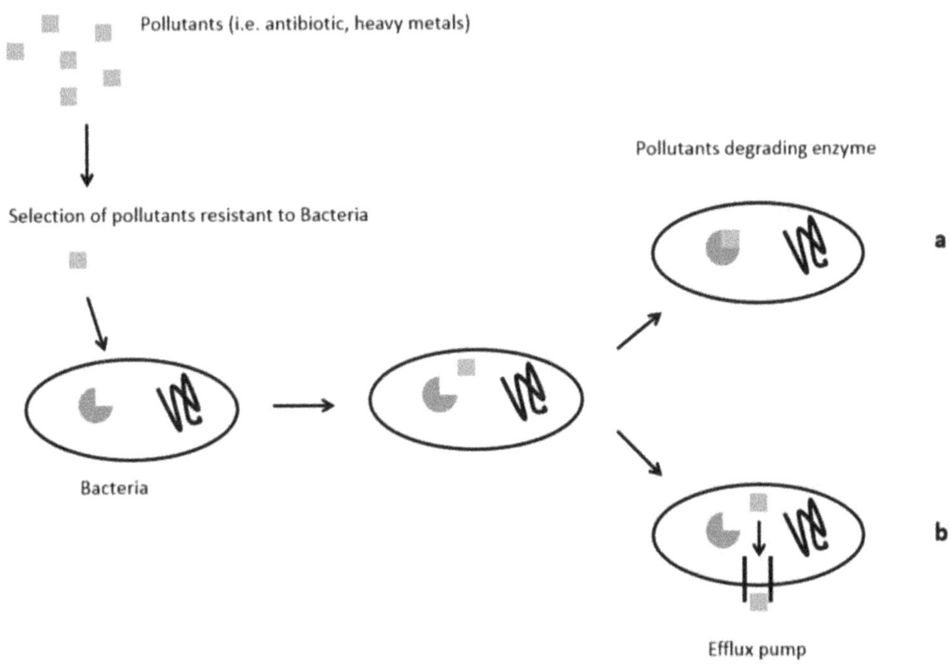

Figure 1 – Pollutants select Bacteria that can adopt different survival strategies. [a] Inactivation of the pollutants via modification or enzyme degradation process. [b] Some Bacteria possess an efflux pump located on the cellular membrane with the ability to extrude pollutants out of the cell (Borruso 2014).

In specific cases, the genes that confer resistance to antibiotic and heavy metals can be integrated in different types of mobile genetic elements named plasmids, transposons and integrons. Plasmids are small, circular and double-stranded molecules of DNA, not essential for cell life, and capable of replicating independently. They are, mainly, responsible for the spreading of antibiotic- and heavy metal-resistant genes among microorganisms (intercellular mobility or horizontal gene transfer), which are often taxonomically distant (Heuer & Smalla 2007). Transposons are segments of DNA that facilitate the transfer from one genetic locus to another one in the same cell (intracellular mobility) or among cells (conjugative transposons) (Hall & Collis 1995). Integrons are genetic elements able to capture, carry and express genes (known as gene cassettes) associated with antibiotic and heavy metal resistance. Integrons are not self-mobilisable, but they are usually located in composite transposons or on plasmids facilitating their mobility (Nemergut et al. 2008).

Borruso (2014) analysed the presence of Class I integron and the associated genes cassettes in the city of Zhangye, Gansu Province, northern China. Sediments associated to *Phragmites australis* roots in freshwater channels characterised by a high level of pollution were studied, and the authors reported the presence of integrons carrying various gene cassettes in all polluted sites. Differently, integrons and relative gene cassettes were not detected in the unpolluted area used as control. These results indicate that Class I integrons could be a promising bioindicator in freshwater environments affected by a broad spectrum of pollutants. As for the Eukaria, microorganisms are extensively used to assess the environmental quality as "tester organisms". The bioluminescent Bacteria *Pseudomonas fluorescens, Vibrio fischeri* and *Vibrio harveyi* have been broadly used to detect the level of xenobiotics such as pesticides, heavy metals or organic compounds in a sample. The light that is normally emitted by the Bacteria decreases in presence of toxic substances, since pollutants inhibit or disrupt the cellular metabolism. This test can be used for analyses of short- and long-term effects of contaminants (Girotti et al. 2008).

The presence and abundance of faecal coliforms are indicators of microbiological water quality, since they are indicative of faecal contaminations and of the possible presence of enteric pathogens. This group includes *Klebsiella* sp., *Escherichia* sp., *Citrobacter* sp. and *Enterobacter* sp., associated to the intestine of warm-blooded animals and easily found in their faeces (APHA 1995). Although extensive research literature has focused on monitoring specific bacterial species or a limited bacterial population to assess the health of freshwater ecosystems, the use of the entire microbial community for environmental monitoring has been receiving attention only recently.

A few works have demonstrated that the entire bacterial community is a promising tool to predict freshwater health status, granted, its sensitivity to the presence of contaminants and environmental stresses. As already stated, bacterial communities have the ability to change their taxonomic and physiological features according to environmental stresses and to contaminants they are exposed to (Lear & Lewis 2009; Sun et al. 2012; Borruso 2014). Biomonitoring at the community level integrates numerous taxa that bioindicate a broader aspect of freshwater environments, underlying the occurrence of different types of disturbance. This approach is more robust because it reflects changes for multiple species, including rare species types. On the contrary, a single taxon, which may be a limited bacterial population, may potentially remain unaffected by the same disturbance. The bacterial community composition of six estuaries, three with a high level of anthropic impact and three less impacted ones, was analysed by Sun et al. (2012). They found a differentiation in the microbial community composition between polluted and moderately polluted samples. The differences among the microbial communities composition in the same-site results were limited, confirming that they do not differ if exposed with same-environmental variables. Similar outcomes were obtained in a study in which the authors found a link between land use and the microbial communities associated to stream sediments. The bacterial community structures analysed in samples collected in the rural and urban area showed striking differences (Lear & Lewis 2009).

3.3 Methods to assess microbial community diversity and structure in freshwater sediments

There is a variety of techniques to study freshwater microbial diversity. Traditional methods are based mainly on culturing methodologies that use a variety of culture media designed to select several different microbial taxa. Culture-based methods are important to isolate and study bacterial strains, but they are not the optimal tool to evaluate the overall microbial diversity, given that conditions they offer are usually selective for a particular population of microorganisms. It is estimated that less than 1 % of the known bacterial species can be isolated by using traditional techniques, since the vast majority of microorganisms are not able to grow due to the lack of required environmental conditions that cannot be simulated in the laboratory (Curtis et al. 2002). Other Bacteria are intrinsically not cultivable, due to physiological constraints such as quorum sensing growth limitations or the necessity to grow in co-culture with other species.

In the last few decades, several biomolecular methods have been developed to study uncultivable microorganisms, allowing a new perspective for the analysis of microbial community diversity and structure. This approach is based on the Polymerase Chain Reaction (PCR). DNA extracted from the sample is analysed to detect microorganisms. The most common approach is the PCR amplification of the 16S rRNA conserved gene. In particular, this gene is broadly used because of its noteworthy features, namely its essential function, evolutionary properties and characteristic of having highly conserved as well as species-conserved regions. Moreover, 16S rRNA gene sequences are commonly used as a housekeeping genetic marker to study bacterial phylogeny and taxonomy, mainly due to the fact that it is present in all Bacteria and the fragment is large enough (1,500 bp) for bioinformatic purposes (Woese 1987; Neefs et al. 1993).

Once DNA has been amplified, a crucial point is the separation of the amplified fragment from the non-target sequences. Different techniques are available for sequence separation. Cloning libraries involve the ligation of the amplified genes into a plasmid vector and the transformation of *Escherichia coli*, followed by the screening of the obtained clones. Alternatively, fragments with the same size but different sequences can be separated via the use of Denaturing Gradient Gel Electrophoresis (DGGE). This technique allows the analysis of different microbial communities simultaneously on the same gel. Sequences are separated because of the different dissociation behaviour of the DNA fragments. After the run, it is possible to cut the single bands and sequence them (Muyzer et al. 1996). The sequence identification is done by comparing them with those of known organisms in a large database such as the Ribosomal Database Project (RDP) (http://rdp.cme.msu.edu/) and GenBank (http://www.ncbi.nlm.nih.gov/).

Electrophoretic profiles can be used to represent the investigated microbial structure by displaying the community in a band or peak profile that can be used for statistical comparison. Each band or peak represents a taxon. This approach is useful to assess microbial communities' differences among samples. If applying automated methods, fragments are marked with a fluorescent chromophore, separated through capillary electrophoresis and detected by a "CCD camera" after being excited by a laser-light. Among these methods, Terminal-Restriction Fragment Length Polymorphism (T-RFLP) implies that the 16S rRNA gene is amplified by using two specific primers with two different fluorophores, and then digested with a restriction enzyme. The laser will detect only the terminal fragments obtained after the restriction. The different sizes of the digested terminal fragments of the 16S rRNA gene represent the different taxa, and in some cases at the genus level (Liu et al. 1997).

In addition, the microbial structure can be studied with another automated method called Automated Ribosomal Intergenic Spacer Analysis (ARISA), based on the investigation of the amplified intergenic region between the 16S and 23S rRNA genes (ITS). Being able to detect differences up to a single-nucleotide, the technique shows a high resolution, up to the sub-species level, and reproducibility is guaranteed by instrumental automatism (Fisher & Triplett 1999; Cardinale et al. 2004). Although, automated techniques can be used to analyse microbial community profiles, offering a huge amount of information with respect to traditional techniques, they do not sufficiently describe microbial diversity in depth. Next generation sequencing is becoming a routinely used technique able to provide deeper insights into complex microbial life.

Pyrosequencing is a flexible, parallel-processing and easily automated method for DNA sequencing. It has a higher throughput and coverage of phylotypes compared to other techniques. One of the primers used to amplify the fragment of interest is modified with biotin. The fragment is mixed with the enzymes DNA polymerase, ATP sulfurylase, luciferase and apyrase, the substrates adenosine-5-phosphosulfate (APS) and luciferin.

Later, the four nucleotides are added one at a time, iteratively, in the nucleic acid polymerisation reaction. Pyrophosphate (PPi) is released during the ATP-conversion operated by ATP sulfurylase, and light is emitted, while luciferin is converted into oxyluciferin; this latter reaction is catalysed by luciferase. The light produced emits a signal, detected by a camera, proportional to the number of nucleotides incorporated during DNA synthesis. The process is repeated with each one of the four nucleotides (dAGP, dGTP, dCTP and dTTP) until the DNA sequence of the single stranded template is synthesised. The sequential collection of images taken by the camera is analysed to measure the light intensity in order to work out the amount of a specific dNTP incorporated in a given attempt. The imagine analysis permits for the calculation of a number of sequences per bead (Margulies et al. 2005; Sogin et al. 2006).

3.4 Bacterial bioindication to assess water quality in different land-use systems: Two case studies in northern China

Northern China has a dry climate, abundant sun radiation, strong winds and little precipitation, concentrated in a restricted period of the year. The territory is characterised by a number of arid biomes, i.e. deserts, semi-deserts, steppes and mountain ecosystems (MWR 2004) – for more topographical information see Chapter 2. In the last 50 years because of the continuously growing population, the rapidly expanding industry and the increase of productive farms, the length of the water channel system has increased by more than 350 % and water demand by more than 250 % (Ringler et al. 2010). Moreover, Fu et al. (2004) reported that climatic change is causing a dramatic decline of the runoff water, followed by a decline of water quality because of the continuous release of several inorganic and organic toxic compounds. Pollution is exacerbated by the loss of cultivable land due to desertification, erosion, salinisation and heavy metal pollution (Kim 2007). The use of fertilisers and pesticides in agriculture has led to an increased presence of nitrogen, phosphate and heavy metals – such as Cd, Pb, Cu and Zn. In addition, the development of metallurgical industries has caused a further escalation in heavy metal pollution (Su et al. 1994; Wang et al. 2001; Cheng 2003) that negatively influences the quality of crops, atmosphere, water and human and animal health (Zhang 1999; Liao 1993).

Zhangye, located in Gansu Province, is situated nearby the Heihe River. The water sources of the Heihe River are the glaciers of Qilianshan Mountain, south of the region. The city is an oasis in an arid region characterised by an urban environment, numerous streams, fertile soil and reed stands. Zhangye is growing in various aspects including mining, production of building materials, electric power, metallurgy, machinery assembly, transportation and agriculture. In recent years, water demand has been dramatically restricted due to the excessive water use in socio-economic systems while environmental pollution has caused a decrease in water quality (SBZC 2003). Moreover, since the water system is not well organised and the irrigation methods are inefficient, desertification is increasingly causing conflicts between local communities (MWR 2004) due to the loss of suitable land (Pei-dong et al. 2007). About 95 % of the water is currently used for agriculture and, in particular, for the cultivation of crops that need a high amount of water.

Borruso (2014) analysed sediment samples associated to the rhizosphere of *P. australis,* in Zhangye, collected in channels exposed to different land uses. Microbial communities, not only resulted in being extremely different in the polluted and unpolluted sites, but they also differed according to the type of pollution (i.e. heavy metals and nutrients; Photograph 1). Furthermore, samples not affected by pollutants showed a bacterial community structure highly similar to the one of the samples collected in a similar natural area in Inner Mongolia (considered as an out-group). The similarity among samples from very distant unpolluted areas of different channels could indicate that, in absence of stressors, the rhizosphere effect is the major driver of bacterial diversity. Therefore, the rhizosphere of *P. australis* can be seen as a normaliser of the bacterial community structure, given that it does not vary between different geographic areas (Figure 2).

The Hetao Irrigation District, located in the western part of Inner Mongolia, has a typical continental climate with moderate precipitation throughout the year and very cold winters and very dry summers. The Hetao Irrigation District hosts the largest farmland drainage and irrigation system of the Yellow River Basin. About 78 % of the water is used for the agriculture in particular for maize, wheat and sunflower cultures (Barton 2005; Fejes et al. 2008). The system is composed of 20,000 branch irrigation channels, which enter into the main drainage channel and finally into Wuliangsuhai Lake (Barton 2005; Fejes et al. 2008). Wuliangsuhai Lake has an area of 33,348 km^2 and a capacity of 2.5×10^8 - 3.0×10^8 m^3. Half of the lake's surface is covered by macrophytes and in particular, *P. australis* as the most dominant species (Barton 2005). The intensive use of fertilisers in the Hetao area has resulted in a large nutrient load in the irrigation water system and eventually in Wuliangsuhai Lake. The fertiliser drained into the Wuliangsuhai Lake increased from 60,000 t in 1980 to 600,000 t in 2000 (Yu et al. 2007). The eutrophication and pollution of Wuliangsuhai Lake is very serious. It is estimated to be 18,750 t y^{-1} of COD, 2,350 t y^{-1} of BOD, 10^6 t y^{-1} of phosphates and 1,673 t y^{-1} of nitrates (Barton 2005; Fejes et al. 2008).

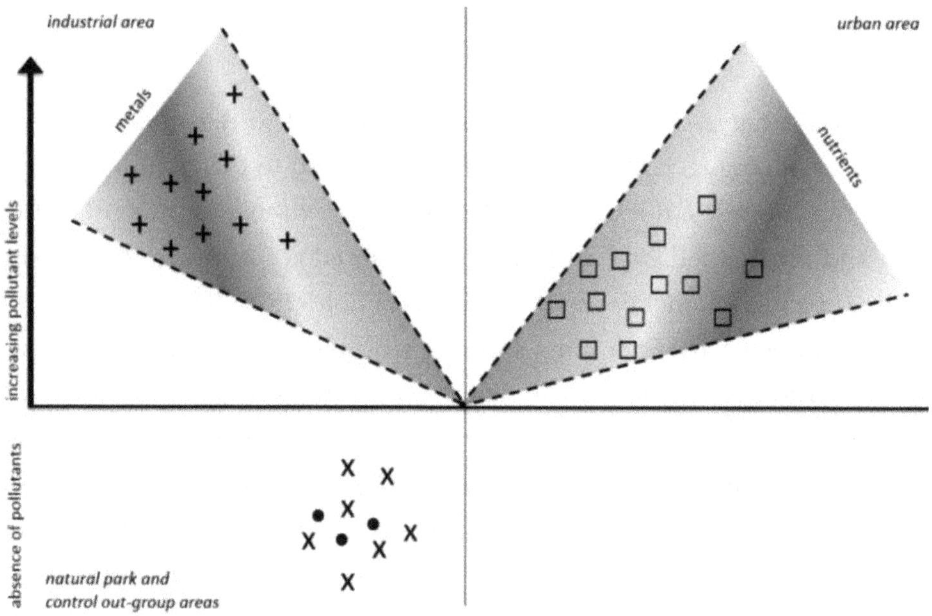

Figure 2 – A multivariate analysis of ARISA profiles via ordination analysis. Metals (Cu, Zn, Pb, Cd, As, Cr, Hg, Mn, Al and Ni) and nutrients (N and P) are influenced by different microbial communities. Industrial area [+], urban area [□], natural park area (x) and out-group [•]; adapted from Borruso (2014).

Sediments associated to *P. australis* were collected from the main drainage of Hetao Irrigation District along a transect of around 250 km. The microbial community structure results differentiated according to a clear biogeographical scale. Samples from the secondary and first part of the main drainage channels of the Hetao tended to group together. The third part of the main drainage channel showed a distinctive bacterial community, probably due to the effect of the entrance of polluted water from Bayannur, urbanised Linhe area, where chemical industries are located. The latter part of the main drainage channel as well as Wuliangsuhai Lake were characterised by very different bacterial communities less influenced by metals and agricultural nutrients. Microbial communities from the samples of the lake outgoing channel showed distinctive profiles, originated from those of the lake. A close relationship between the microbial community structures analysed by ARISA and the geography of the sampling sites was found. Accordingly, we found that water characterised by the observed chemical pollution or by supposed organic pollution clustered differently from those in areas that are more natural.

3.5 Conclusion

The remarkable developments made within biomolecular sequencing techniques and innovative fingerprinting analysis has allowed microbiologists to deeper analyse environmental samples. They can now overcome problems arising from bacterial strains which are not unculturable in a laboratory environment. Microbial bioindicators, not excluding other novel available methodologies, are a useful tool in context, characterised by environmental factors that cannot be directly measured. Examples include compounds derived by pesticides or toxic waste containing a number of interacting pollutants. Microbial communities should be considered as one of the first environmental parameters to monitor in order to have a fast response into ecosystem health and its relating factors.

Key references

Borruso, L. 2014. Rhizobacterial communities as bioindicators of environmental stresses in freshwater ecosystems. PhD Thesis. Bolzano, Italy: Free University of Bozen-Bolzano.

Girotti, S., Ferri, E.N., Fumo, M.G. & Maiolini, E. 2008. Monitoring of environmental pollutants by bioluminescent bacteria. Analytica Chimica Acta, 608(1): 2–29.

Kumar, M. & Khanna, S. 2010. Diversity of 16S rRNA and dioxygenase genes detected in coal-tar-contaminated site undergoing active bioremediation. Journal of Applied Microbiology, 108(4): 1252–1262.

Lear, G. & Lewis, G.D. 2009. Impact of catchment land use on bacterial communities within stream biofilms. Ecological Indicators, 9(5): 848–855.

McGeoch, M.A. 1998. The selection, testing and application of terrestrial insects as bioindicators. Biological Reviews of the Cambridge Philosophical Society, 73(2): 181–201.

Piddock, L.J. V. 2006. Multidrug-resistance efflux pumps - not just for resistance. Nature reviews. Microbiology, 4(8): 629–636.

Stepanauskas, R., Glenn, T.C., Jagoe, C.H., Tuckfield, R.C., Lindell, A.H., King, C.J. & McArthur, J. V. 2006. Coselection for microbial resistance to metals and antibiotics in freshwater microcosms. Environmental Microbiology, 8(9): 1510–1514.

Sun, M.Y., Dafforn, K.A., Brown, M. V. & Johnston, E.L. 2012. Bacterial communities are sensitive indicators of contaminant stress. Marine Pollution Bulletin, 64(5): 1029–1038.

4. Water allocation along the Heihe River and Tugai forest conservation in Ejina

Niels Thevs
Henrike Hochmuth
Jan Felix Köbbing

4.1 Introduction

Tugai forests are the riparian forests of the deserts of Central Asia (Ogar 2003; Treshkin 2001). These forests together with reed beds are the most productive ecosystems of the deserts of Central Asia (Thevs 2007; Thevs et al. 2012) and harbour most of the biodiversity of these deserts (Thevs 2007; Thevs et al. 2008b; Ogar 2003). In the past, Tugai forests were a major wood source and an important pasture in the deserts of Central Asia (Hoppe 1992). During the past five decades, their role to stabilise moving sand has been acknowledged (Song et al. 2000; Yimit et al. 2006). Today, their role with respect to carbon sequestration and groundwater recharge has been attracting more and more attention (Thevs et al. 2012). Tugai forests play an important role with regard to the identity of people in these desert regions (Hoppe 1992; Ejina Qizhe 1998). During the past decade, Tugai forests in Xinjiang and along the Heihe River attract more and more tourists, with Ejina being China's most important tourist destination with respect to Tugai forests (personal observation).

From east to west, these riparian forests naturally are distributed along rivers in northwest China (Inner Mongolia, Gansu and Xinjiang), Mongolia, low-lying river valleys of Kyrgyzstan and Tadzhikistan and in the deserts of Kazakhstan, Uzbekistan and Turkmenistan (Lavrenko 1956; Wang et al. 1996; Ogar 2003). The largest Tugai forests were distributed along the Amu Darya (within Turkmenistan and Uzbekistan) and in the Tarim Basin, China with approximately 500,000 ha each in the 1950s (Huang 1986; Treshkin 2001). In China, after the Tarim Basin the second largest Tugai forests naturally were distributed along the downstream section and the delta of the Heihe River in Ejina County, Inner Mongolia. In Ejina, about 430,000 ha are covered with woodland, including Tugai forests (Ejina Qizhe 1998). The Tugai forests in Ejina were described as a green belt in the middle of the desert as early as 1927 by Sven Hedin, who crossed the Heihe River during the Sino-Swedish Expeditions between 1927 and 1935 (Hedin 1943).

After the People's Republic of China was founded in 1949, oases areas along all rivers in northwest China were enlarged. As agriculture in northwest China largely depends on irrigation, more and more water was diverted from the rivers in Xinjiang, Gansu, Inner Mongolia and other provinces in the north western parts of the country. The Heihe River frequently ceased to reach its two end-lakes, West and East Juyanhai Lakes, so that the West Juyanhai Lake fell dry in 1961, while the East Juyanhai Lake shrunk and fell dry for the first time in 1973 (Ejina Qizhe 1998). Agriculture and animal herding declined, too, due to severe water shortage, resulting in widespread poverty in Ejina County.

In 2000, a water allocation plan was adopted by the Central Government of China, which should ensure a guaranteed annual amount of 0.95 km^3 to be released into Ejina County. One objective of this water allocation plan was to restore the Tugai forests along the Heihe in Ejina. Against this background, we will analyse the current state of the Tugai forests along the Heihe in Ejina and draw conclusions in how far the water allocation plan meets the needs for Tugai forest conservation. We will analyse the state of the Tugai forests according to the criteria for forest quality by IUCN and WWF in 1996 (WWF and IUCN 1996). The assessment of the forests is based on two field visits in 2011 and 2012.

4.2 Ecology of Tugai Forests

Tugai forests are the riparian forests distributed along the rivers, which flow through the deserts of Central Asia, e.g. Amu Darya, Syr Darya, Chu, Ili,

Irtysh, Tarim and Heihe (Wang et al. 1996; Treshkin 2001; Ogar 2003). The Tugai forests are formed by the Poplar species *Populus euphratica* and *P. pruinosa*, willow species, like *Salix acmophylla* and *S. soongorica*, and *Elaeagnus angustifolia* (CMF 1990; Wang et al. 1996; Ogar 2003). The understory vegetation is dominated either by shrubs, i.e. *Tamarix* species, *Halimodendron halodendron*, *Lycium ruthenicum*, *Nitraria sibirica* and partly by halophytes, or by *Phragmites australis* and herbs like *Glycyrhizza glabra* and *Alhagi sparsifolia* (Wang et al. 1996; Thevs et al. 2008a). *Salix* species are restricted to Tugai forests along the rivers in the Zhunggar Basin (Xinjiang), Ili Basin and the Aral Sea Basin (CMF 1990; Wang et al. 1996; Ogar 2003). *H. halodendron* is much more frequent along the Syr Darya and Ili compared to the Tarim and Heihe. In addition, the herb layer is more diverse in the Zhunggar Basin (Xinjiang), Ili Basin and the Aral Sea Basin compared with the Tarim Basin and the Heihe River Basin (Wang et al. 1996).

P. pruinosa is distributed throughout the Tugai forest area, but it is restricted to more humid, less saline, and less winter-cold sites compared to *P. euphratica*. Thus, *P. pruinosa* is neither found in the downstream region of the Amu Darya and Syr Darya, nor in the eastern part of the Tarim Basin, nor in the northern part of the Heihe River Basin, which corresponds to Ejina County (Wang et al. 1996). In all these three regions without *P. pruinosa*, the Tugai forests are built by *P. euphratica* with some minor stands or individuals of *E. angustifolia* (Huang 1986; Wang et al. 1996; Ejina Qizhe 1998; Treshkin 2001; Thevs et al. 2008a). In these three regions, most Tugai forests have only sparse or do not have any understory vegetation. Only on sites, which are submerged during flood events or which are located very close to river courses, understory vegetation is found.

Tugai forests in Ejina thus practically are *P. euphratica* forests without undergrowth or in parts with understory vegetation of mainly *Tamarix* species and *P. australis*. The plants of these riparian ecosystems survive under the arid climate, because they take up water from the groundwater as obligate or facultative phreatophytes (Sukhova & Gladyshev 1980; Huang 1986; Xinjiang Linkeyuan Zaolin Zhisha Yanjiusuo 1989; Ogar 2003; Rüger et al. 2005; Thevs et al. 2007, 2008a). *P. euphratica* and *P. australis* are obligate phreatophytes, i.e. these species must have continuous contact to the groundwater (Gries et al. 2003; Thomas et al. 2006). In contrast, *Tamarix* species are facultative phreatophytes and thus are able to survive a certain period disconnected from the groundwater, but using soil moisture from the unsaturated zone (Smith et al. 1998).

Once established, *P. euphratica* can grow on sites with the groundwater level as deep as 12 m (Kuzmina & Treshkin 1997; Novikova 2001; Thevs 2007; Thevs et al. 2008a). *Tamarix ramosissima*, one of the most widely distributed *Tamarix* species, also grows on groundwater level as deep as 12 m (Thevs et al. 2008a). *P. australis* is distributed on sites with groundwater levels not deeper than 3 m (Huang 1986; Liu et al. 1990; Novikova 2001; Thevs 2007). The majority of other species, which may occur in the undergrowth of Tugai forests, is restricted to sites with groundwater levels not deeper than 5 m (Thevs et al. 2008a). The groundwater is replenished by the river courses; therefore, the groundwater levels drop when moving from the riverbanks away from the river (Hou et al. 2007). According to deeper and deeper groundwater levels, the number of plant species, which occur in a Tugai forest, decreases with increasing distance from the river course (Thevs et al. 2008a).

With regard to recruitment, *P. euphratica* follows two strategies, i.e. generative and vegetative recruitment. Generative recruitment depends on flood events and river dynamics (Wang et al. 1996; Thevs et al. 2008a; Wiehle et al. 2009; Eusemann et al. 2013). The seeds are light, have pappus-like hairs, and are dispersed by wind and water. The main fruiting period is between July and September, i.e. during the flood period of the rivers in the distribution area of *P. euphratica*. Optimal germination occurs under conditions of intensive sun radiation, a temperature between 25 and 30 °C, and water-saturated soils with a salt content lower than 0.2 %. Such sites are created by the annual summer floods and shifting river courses. Germination occurs in lines or narrow strips marking flood water lines on the riverbanks (CMF 1990; Liu et al. 1990; Eusemann et al. 2013). After germination, the seedlings invest more in root growth than in shoot growth in order to secure water uptake during the spring and early summer of the following year.

During spring of the year following a germination event, the groundwater level under the seedlings and soil moisture in the root zone of the seedlings drop. Often, the seedlings die, because the roots lose contact to the groundwater and cannot take up enough water from the dwindling soil moisture. Therefore, it is crucial that in the year after germination there is another flood event, which starts in time and reaches a water level so high that the groundwater is lifted into the root zone of the seedlings, but not too high that the seedlings are not drowned. Once the roots of a seedling have grown so deep that they tap the groundwater under the riverbank the whole year round, we call the seedling established. Most likely, a heterogeneous soil structure of sand, silt, and thin silty-clayey soil horizons helps seedlings to survive, because the silt and silty-clayey soil horizons keep more plant available soil moisture than sand so that the seedlings may take up water during spring of the year after germination. The silty-clayey soil horizons must not be too thick or too clayey, because the seedlings' roots cannot penetrate such horizons (Thevs et al. 2008a). After a seedling is established, its root system develops horizontal roots, from which root suckers emerge.

On sites out of the reach of the floods, *P. euphratica* only is able to recruit from root suckers, which emerge from the lateral roots, i.e. through clonal reproduction. These clones can cover areas of 4 ha (Bruelheide et al. 2004). As the parent trees supply the root suckers with water until they reach the groundwater by themselves, vegetative recruitment is restricted to sites on which *P. euphratica* does not suffer water stress. In general, these are sites with a groundwater level not deeper than 6 m (Thevs et al. 2008a).

Under natural conditions, which include regular flood events and river dynamics, the life cycle of a *P. euphratica* Tugai forest can be characterised as follows (Thevs et al. 2008a, 2008b; Wiehle et al. 2009; Eusemann et al. 2013): *P. euphratica* and other species germinate at a river bank after a flood event. *P. euphratica* and other species are able to establish. It also forms larger and contiguous forests with its root suckers. Other species form the undergrowth. As the river dynamics continues to relocate the river branch, at which bank this forest has formed, the river branch may either erode the area of this forest site or it may move away from this forest site. If the river branch moves away from that forest site, the groundwater level will gradually drop. The plant species have to grow deeper and deeper roots following the groundwater. According to their respective ability to do so, the plant species will survive or disappear from this particular forest site. In the course of dropping groundwater levels, the salt content of the groundwater may increase. Increasing salinity results in more species that disappear from this forest site. Therefore, within the succession of Tugai forests species disappear but no new species can enter the succession. Along old and permanently dry river branches, pure *P. euphratica* Tugai forests are distributed without any other species as undergrowth. Under more saline conditions, *P. euphratica* disappears and only *Tamarix* and halophytes may remain.

P. euphratica forests have an annual biomass increase of aboveground woody biomass of up to 1.5 t/ha * a (2.5 t/ha * a on extremely productive sites), as calculated on the basis of tree ring widths (Xinjiang Linkeyuan Zaolin Zhisha Yanjiusuo 1989). Thevs et al. (2012) found biomass stocks (above- and belowground) of up to 44 and 58 t/ha and annual increments of 1.3 and 2.6 t/ha * a in the Huyanglin Nature Reserve at the Tarim River and the Amu Darya State Reserve in Turkmenistan, respectively. The site at the Tarim River has a groundwater level of 3.5 m below surface and a tree density of 379 trees per hectare with an average DBH of 20.2 ± 13.8 cm. The respective site in the Amu Darya State Reserve has a groundwater level of 1.9 m below surface and a tree density of 964 trees per hectare. The average DBH was 14.1 ± 6.4 cm. The annual water consumption of *P. euphratica* forests like at the former and latter site are 554 – 725 mm (Thevs et al. 2013) and 907 – 1043 mm (Thevs et al. 2014), respectively. The annual water consumption of *P. euphratica* on a site at the Heihe in Ejina, whose average age is 25 years with a canopy density of 0.8 %, an average tree height of 10 m, and average diameter at breast height of 12 cm, was 447 mm (Hou et al. 2010).

4.3 Water allocation along the Heihe River

The Heihe River Basin, with an area of 120,000 km^2, is the second largest endorheic river basin of China (Li et al. 2012b). The headwaters of the Heihe River are located in the Qilian Mountains in Gansu Province south of the city of Zhangye. The Heihe ended in the two terminal lakes West and East Juyanhai close to the border of Mongolia (Figure 1, Chapter 2). At the gauging station Yingluoxia, the Heihe flows out of the Qilian Mountains into the oasis of Zhangye with its population of about 1.3 million people. Within the area of Zhangye, about thirty small rivers flow down from the Qilian Mountains, which now are diverted into irrigation. Only during spring or high flood events some of these rivers reach the Heihe. Once the Heihe leaves Zhangye, it flows as a so-called losing stream through mainly gravel deserts northwards from Gansu into Inner Mongolia. Downstream of Zhangye the Heihe passes the gauging station Zhengyixia. From this station, the Heihe passes the small oasis Jinta and flows into Ejina County in Inner Mongolia. About half way between Zhengyixia and the terminal lakes there is the gauging station Langxinshan. There, the Heihe splits into two branches flowing into West and East Juyanhai Lake, respectively (Qi & Luo 2005) as shown in Figure 1, Chapter 2. Until the 1960s, the two branches branched off further and formed an inland delta.

Table 1 – Aggregated climatic data of Zhangye and Ejina from 1973 to 2012, sourced from TuTiempo.net.

Climate station	Zhangye	Ejina
Position	38.93 °N, 100.43 °E	41.95 °N, 101.61 °E
Elevation [m a.s.l.]	1,483	941
Annual mean temperature [°C]	8	9.4
January mean temperature [°C]	-9	-10.7
July mean temperature [°C]	22,1	27.2
Annual precipitation [mm]	170	60

The climate in the Heihe River Basin is arid and continental, as shown in Table 1. In the Qilian Mountains along the headwaters of the Heihe River, the annual precipitation is about 400 mm. At the foothills of the Qilian Mountains, i.e. climate station Zhangye, it drops to 170 mm and further north in Ejina it decreases to 60 mm (Table 1). About two thirds of the annual precipitation is concentrated in the months June to August. This precipitation maximum falls together with the snow and glacier melting period in the Qilian Mountains, which results in annual summer floods in the Heihe River (Figure 1). Table 2 shows the development of cropland along the Heihe from the 1970s to today. The composition of crops in Zhangye is given in Table 3. Cropland also was reclaimed in Ejina County.

Table 2 – Area of cropland [ha], including fallow and planted forests, digitised from Landsat satellite images.

Administrative unit	Area of cropland [ha], including fallow and planted forests in the year			
	1975	1990	2000	2010
Zhangye	172,388	283,827	323,240	369,336
Jinta	14,996	28,504	42,024	56,826
Jiayuguan	40,749	67,952	105,540	117,726
Ejina	1,540	3,454	3,040	9,130

This increase of cropland area resulted in an increasing demand for irrigation water, which was diverted from the Heihe River. The annual runoff at the gauging station Zhengyixia decreased from 1.19 km^3 in the 1950s over 0.942 km^3 in the 1980s to 0.475 km^3 between 1990 and 1995 (Feng & Cheng 1998). Thus, the Heihe downstream of Zhengyixia faced severe water shortage. The terminal lake West Juyanhai has been dry since 1961 and most of the western branch of the Heihe in Ejina has been dry, too. The terminal lake East Juyanhai Lake covered 35.5 km^2 in 1958, shrunk to 23.6 km^2 in 1980, and dried up in the beginning of the 1990s (Ejina Qizhe 1998). In 2002, it reappeared for a few months with an area of 12 km^2 (Wang et al. 2002). In the course of decreasing runoff reaching Ejina County, soil salinisation increased in parts of the county (Qi & Cai 2007). Groundwater levels dropped from 0.5 – 1.3 m in the 1940s to 3 – 6 m in the 1990s (Guo et al. 2009). However, the runoff from the Qilian Mountains, recorded at the gauging station Yingluoxia, has not changed significantly for the past 50 years despite of global warming and shrinking glaciers in other mountains in China and Central Asia (Jiang & Liu 2010).

Table 3 – Composition of crops and area per crop in Zhangye in 2012, sourced directly from the Agriculture Administration, Zhangye.

Crop	Area [ha]	Crop	Area [ha]
Seed corn	66,700	Pear and apple	5,300
Potatoes	26,700	Apricot	3,300
Vegetables	26,700	Grapes (for wine)	3,300
Rape	26,700	Cotton	2,000
Wheat	20,000	Grapes (for raisins)	2,000
Dates	10,000	Peach	700

In the course of increasing water shortage along the Heihe in Ejina, the natural riparian ecosystems were severely degraded and rural livelihoods were affected (Jin et al. 2010). The area covered by *P. euphratica* forests was 50,000 ha in the 1950s, shrunk to 22,667 ha in 1998, and increased to 38,663 ha by 2004 due to restoration measures (Lu et al. 2007; Bai et al. 2008). During the 1990s, this area of degraded vegetation was made responsible for dust storms, which affected the region, but also other parts of China (Feng & Cheng 1998). Especially the Tugai forests were considered able to significantly reduce dust storms in Ejina proper and beyond, if restored (Guo et al. 2009).

In 2000, an integrated water resource management of the Heihe River Basin was established. In the frame of this integrated water resource management a water allocation plan between the middle and the lower reaches, i.e. between Zhangye and downstream of Zhangye, was adopted. This water allocation plan starts with an average runoff of 1.58 km^3/a at Yingluoxia. An annual runoff of 0.95 km^3 must pass Zhengyixia gauging station and flow into the lower reaches. According to Zhang et al. (2011), the average annual runoff from 2000 to 2009 was 0.995 km³ at Zhengyixia and 0.53 km³ at Langxinshan. The annual runoff of 0.53 km³ at Langxinshan, which is equivalent to the amount of water that Ejina receives, lies above the annual evapotranspiration of the whole cropland and riparian ecosystems within Ejina. More water was guided into the eastern river branch of the Heihe compared to the western branch so that the reed and shrub vegetation around the East Juyanhai Lake started to recover (Guo et al. 2009). While before 2000 the major crops along the Heihe were cotton and paddy rice, now the major crop is seed corn. Most of the seeds, which are used to crop corn in China, are produced in Zhangye. The water consumption of corn in Zhangye is well within the range of other measurements from China. The further crops in Zhangye are listed in Table 3. This shift of crops away from the water demanding paddy rice and cotton enabled to attain a runoff at Zhengyixia according to the water allocation plan. However, the annual amounts of water, which pass Zhengyixia and Langxinshan, fulfil the requirements of the water allocation plan; the runoff distribution within the years has been changed compared to natural conditions. This is shown in Figure 1.

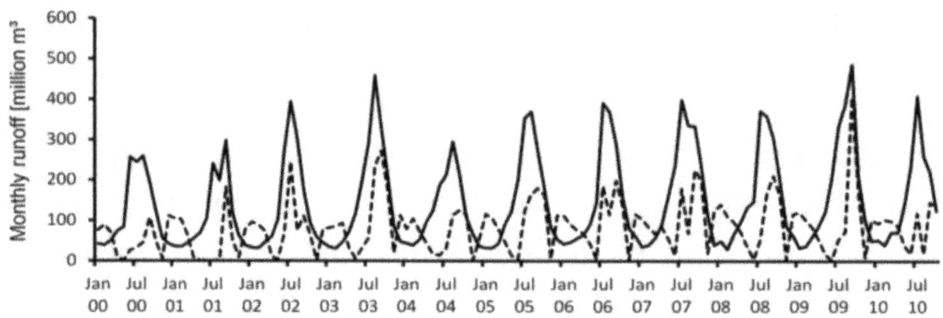

Figure 1 – Monthly runoff of the Heihe River at the gauging stations Yingluoxia (solid line) and Zhengyixia (dashed line), sourced from the Environmental Protection Administration, Zhangye.

While the runoff at Yingluoxia shows clear peaks during the summer months, at Zhengyixia there are two runoff peaks, one in late summer/autumn and another one in winter/early spring. During summer, when the agriculture in Zhangye and other oases has the highest demand for irrigation water, the irrigation demand is covered with water from the Heihe and no water is left for the lower reaches. In autumn and winter/early spring, when irrigation is scaled back and irrigation has not started yet, respectively, water is not needed by the oases along the middle reaches so that water is released into the lower reaches.

4.4 Quality of the Tugai forests in Ejina

The IUCN and WWF developed a set of criteria (Table 4) to evaluate the quality of forests (WWF and IUCN 1996). In this section, we will apply these criteria to assess the quality of the Tugai forests in Ejina against the background of the water allocation plan from 2000. We refer to our own field observations and literature.

Table 4 – IUCN and WWF criteria for forest quality (WWF and IUCN 1996).

Authenticity	Forest health	Environmental benefits	Other social and economic benefits
- natural composition of trees and other species - natural spatial variation of trees with respect to age, size, and proportion of dead timber - continuity - accommodation of natural disturbance patterns within forest management - integration of forest into the landscape - management practices which mimic natural ecological processes	- impacts of air pollutants - robustness to global climate change	- biodiversity conservation - soil and watershed protection - local climatic effects -carbon sequestration	- timber products - non-timber products - support for local industries - recreational value - forest as homeland for people - aesthetic values - historical values - cultural values - educational values - spiritual values - local distinctiveness

4.4.1 Authenticity

The natural tree species composition of the Tugai forests in Ejina consists of *P. euphratica* as nearly the only forest building tree (Photograph 2 to 5) with some stands or individuals of *E. angustifolia*. The Tugai forests today meet this species composition. Furthermore, the forests are associated with *Tamarix* sp. and *P. australis,* which correspond to the natural species composition. Under *P. euphratica* on sites within a distance of 500 m from the two river branches of the Heihe in Ejina, the groundwater levels now are in the range of 1.5 – 3.85 m below surface (He & Zhao 2006; Liu et al. 2007a; Zhu et al. 2009). These groundwater levels fully support the growth and clonal reproduction of *P. euphratica* as well as *Tamarix* and most other potential understory plant species (Thevs et al. 2008a). Only at the downstream section of the eastern branch, the groundwater level was 4.7 m (Guo et al. 2009), which still well sustains *P. euphratica* as well as *Tamarix* (Thevs et al. 2008a). From 2000 to 2005, within 500 m of the river branches the annual increments of *P. euphratica* increased significantly compared to before the year 2000 (Guo et al. 2009), reflecting that groundwater conditions are sufficient to sustain the existing Tugai forests.

Natural stands of *P. euphratica* had a mean age of 141 years, with a minimum of 60 years and maximum of 300 years. The mean tree density was 380 trees per ha. Only planted and semi-natural stands had mean ages of 29 years and 26 years, respectively (Li et al. 2010a). Thus, comparing these natural stands of *P. euphratica* with natural stands along the Tarim River (Thevs et al. 2012) the natural Tugai forests in Ejina are over-aged. By now, natural rejuvenation through root suckers occurs in some *P. euphratica* stands (Photograph 2). Dead timber is present in these natural stands (Photograph 3).

P. euphratica is distributed along both river branches in Ejina within 500 m away from the river branches (He & Zhao 2006). Nevertheless, the Tugai forests do not cover completely these 500 m belts at both sides of the river branches. The Tugai forests are distributed like elongated islands along the river branches. Contiguous *P. euphratica* stands are present and well integrated into the landscape (Photograph 4 and 5). However, small river branches of the previous inland delta of the Heihe in Ejina do not carry water. Therefore, the size and continuity of the Tugai forests lag behind natural conditions (Guo et al. 2009).

The utmost important natural disturbance patterns for Tugai forests are flood events during summer and river dynamics, because they are crucial drivers for the recruitment and development of Tugai forests. Flood events during summer have not occurred since adopting the water allocation plan. Instead, the runoff into Ejina is in autumn and winter. Photograph 2 shows a site, which had been flooded during the previous winter. Liu et al. (2007) found that the water supply in winter has negative impact on annual increments of *P. euphratica* in Ejina.

The two river branches of the Heihe in Ejina have also not been reworked by river dynamics since the year 2000. Summer floods and river dynamics are not mimicked through forest management measures. Therefore, the Tugai forests in Ejina lack generative reproduction and thus conservation of genetic diversity. Artificial rejuvenation of *P. euphratica* has been taken place during the past four decades (Li et al. 2010a), probably with root suckers. However, in how far genetically diverse root suckers have used is not known to us.

4.4.2 Forest health

Air pollution from industry does not occur in Ejina, which is a remote rural community. Dust may play a role, as it covers the leaves and may reduce solar radiation. However, the importance of dust as factor, which may impair the growth rates of *P. euphratica*, lags far behind sinking groundwater and groundwater salinisation.

The robustness or vulnerability to climate change of the Tugai forests in Ejina are decided in the headwaters of the Heihe River in the Qilian Mountains. If climate change altered the runoff regime of the Heihe, it would affect the Tugai forests, because their water supply source exclusively is groundwater, which is delivered by the Heihe. The runoff from the Qilian

Mountains, recorded at the gauging station Yingluoxia, has not changed significantly for the past 50 years despite of global warming and shrinking glaciers as reported from other mountains in China and Central Asia (Jiang & Liu 2010).

4.4.3 Environmental benefits

The two criteria environmental benefits and other social and economic benefits reflect the ecosystem services as widely accepted today (MEA 2005; TEEB 2010). Environmental benefits by WWF and IUCN (1996) refer to regulatory and supporting ecosystem services as understood by MEA (2005). Tugai forests in Ejina provide habitat for wildlife (Ejina Qizhe 1998) as well as Tugai forests all over their distribution area (Wang et al. 1996). The Tugai forests in Ejina are strictly protected and large parts of the forests close to the river branches are fenced. Unfenced forested areas offer unrestricted habitat for birds and other wildlife and decrease environmental pressures from human intrusion. The natural and the planted *P. euphratica* stands play an important role to fix sand and thus reduce dust pollution and improve the local climate. Therefore, the Tugai forests in Ejina contribute significantly to soil protection.

Tugai forests in general are among the most productive ecosystems in the deserts of Central Asia (Thevs et al. 2012). Due to the arid climate, deadwood presumably decays slowly so that Tugai forests should sequester some carbon. However, in how far carbon remains in such ecosystems has not been systematically investigated yet (Thevs et al. 2011). On a local scale, they thus may play a role with respect to carbon sequestration, but due to their small total area, their global role is very limited.

4.4.4 Other social and economic benefits

Other social and economic benefits refer to provisioning and cultural ecosystem services (MEA 2005). The provisioning and cultural ecosystem services of the Tugai forests in Ejina centre on recreation and tourism. Ejina has become a major tourist destination of northwest China with almost 200,000 visitors per year. The visitors travel to Ejina in September and October, in order to see the Tugai forests in autumn due to their aesthetic value and their local distinctiveness. The Tugai forests in Ejina are the most visited Tugai forests in China. Tourists often combine a visit to Ejina with visits to the ancient city Heicheng and the western end of the Chinese wall in Jiayuguan. In this respect, to some extent historical values are attributed to the Tugai forests as part of an area with historical value.

The vegetation and water resources along the Heihe before and after adopting the water allocation plan have been intensively researched by Chinese and some international scholars. Therefore, the Heihe, including the Tugai forests in Ejina, is well known to the scientific community and partly beyond as a showcase for a water allocation plan in a closed river basin. In this context, the Tugai forests contribute to the educational value of the whole region. Today, the Tugai forests are strictly protected so that there are no people living in these forests and no products are made from the timber. Additionally, *P. euphratica* wood has a poor quality as timber as well as fuel wood so that in northwest China *Populus alba* is much more attractive as timber.

4.5 Conclusion

The Tugai forests in Ejina under the current water management serve as a basis for tourism and recreation of significance for whole China and thus create income for people in Ejina. The forests help improving the local climate through sand fixation and reduction of dust storms. Furthermore, the Tugai forests presumably may sequester carbon. The natural species composition is present under the current water management. The annual increments have increased and in some areas natural rejuvenation has started so that the age structure slowly changes from over-aged to diverse. However, the rejuvenation is restricted to clonal reproduction. Under the current water management, the genetic diversity cannot be protected on the long run. Either more water must be released from Zhangye in summer during peak demand of irrigation or seedlings must be grown artificially and planted along the Heihe.

Key references

Ejina Qizhe. 1998. Ejina Qizhe (Description of Ejina County). Beijing: Fangzhe Chubanshe.

Guo, Q., Feng, Q. & Li, J. 2009. Environmental changes after ecological water conveyance in the lower reaches of Heihe River, northwest China. Environmental Geology, 58(7): 1387–1396.

MEA. 2005. Ecosystems and Human Well-Being: Wetlands and Water Synthesis. World Resources Institute ed. Washington, DC: Millennium Ecosystem Assessment.

Ogar, N.P.P. 2003. Vegetation of river valleys. In: E. I. Rachkovskaya, E. A. Volkova, & V. N. Khramtsov eds. Botanical geography of Kazakhstan and middle Asia (Desert region). Komarov Botanical Institute of Russian Academy of Sciences. Saint Petersburg, Institute of Botany and Phytointroduction of Ministry of Education and Science of Republic Kazakhstan. Almaty, Institute of Botany of Academy of Sciences of Republik Uzbekistan, 313–339.

Thevs, N., Buras, A., Zerbe, S., Kühnel, E., Abdusalih, N. & Ovezberdiyeva, A. 2012. Structure and wood biomass of near-natural floodplain forests along the Central Asian rivers Tarim and Amu Darya. Forestry, 85(2): 193–202.

Thevs, N., Zerbe, S., Schnittler, M., Abdusalih, N. & Succow, M. 2008b. Structure, reproduction and flood-induced dynamics of riparian Tugai forests at the Tarim River in Xinjiang, NW China. Forestry, 81(1): 45–57.

Treshkin, S.Y. 2001. The Tugai Forests of Floodplain of the Amudarya River: Ecology, Dynamics and their Conservation. In: S. W. Breckle, M. Veste, & W. Wucherer eds. Sustainable Land-use in Deserts. Heidelberg, Germany: Springer, 95–102.

Wang, S.J., Chen, B.H. & Li, H.Q. 1996. Euphrates Poplar Forest. Beijing: China Environmental Science Press.

WWF and IUCN. 1996. Forests for Life - The WWF/IUCN forest policy book. Surrey: WWF-UK.

5. Stoichiometry and functional traits of reed (*Phragmites australis*)

Liping Li
Stefan Zerbe
Ping He
Niels Thevs
Giuseppe Tommaso Cirella
Jan Felix Köbbing

5.1 Introduction

Functional traits and life histories of plant species as well as the structure and function of ecosystems are of general and fundamental interest in ecology (Kerkhoff et al. 2006; McGill et al. 2006; Pérez-Harguindeguy et al. 2013). Plants grow in varying environments and there is a challenge for these organisms to adapt to extreme habitats such as very wet environments (e.g. floodplains, lakes and mires) or extremely dry ones (e.g. deserts and semi-deserts). There are many studies on functional traits available on terrestrial plants (Reich & Oleksyn 2004; Han et al. 2011), but only a very few on aquatic ones. Hydrophytes, for example, grow in the water where a limitation of oxygen availability is given. Thus, they require the development of certain tissue characteristics in order to combat this limitation.

Leaf and root traits are important indicators for plant strategies as they have an associated functional purpose (Eissenstat & Yanai 1997). The root traits are closely associated with the nutrient availability of the environment and the plant growth rates (Grime 1977). Plants with high growth rates usually have a high specific root length (SRL), low root diameter, low tissue density, high plasticity and are more adapted to nutrient rich environments. On the contrary, species with a low growth rate are found on nutrient poor sites and have a high root:shoot ratio (RS) (Chapin et al. 2002). Leaves with a low specific leaf area (SLA) have a high dry matter content (DMC), high longevity, high nutrient-use efficiency and better defence against herbivores and physical hazards (Baruch 2011). While leaves with a high SLA have a high photosynthesis and growth rate, correspondingly they have low longevity and less developed defence systems (Baruch 2011; Shi et al. 2010). Factors such as top-soil freezing in winter and nutrient limitations determine the relationships among leaf and root biomass and nutrient content (Craine et al. 2005). Leaves and roots might also be functionally associated with other organs, for example, plant stems (Freschet et al. 2010).

In our investigation with regard to stoichiometry and functional traits, we choose reed (*Phragmites australis*) as a focal species. Reed can cope with a broad range of ground water levels, nutrient supplies, soil and water salinity and land-use (Zerbe & Thevs 2011; Taisan 2009). Reed not only has important ecosystem functions, but provides also a broad range of ecosystem services, such as building material, fodder for grazing, energy source and water purification (Köbbing et al. 2013; Patuzzi et al. 2013a). We examined the stoichiometry, functional traits and community structure of common reed in two settlements of northern China, differentiating between terrestrial and aquatic habitats. Thus, this research also contributes to the knowledge of nutrient content of water and sediment in both Wuliangsuhai Lake and the wetlands of Zhangye. The potential of N and P removal by reed with regard to wetland restoration is discussed. The findings of this study could have implications for the management of reed-dominated wetland ecosystems by understanding trade-offs of plant functional traits in changing climatic conditions. Climate change will become increasingly an issue especially in high latitude areas, like in northern China, where there has been a strong warming effect in recent years (Piao et al. 2010). Water plants are more sensitive to the impacts of climatic change than terrestrial plants. Recent research stresses a significant loss of species and alteration of community structure of wetlands in the last four decades (Fang et al. 2006). To better understand this impact, differences in the responses of functional traits of water and terrestrial plants to environmental change are important for predictive and scenario-based action. *P. australis* could be a model plant to exemplify wetland ecosystem changes in a dynamic global climate.

5.2 Materials and methods

5.2.1 Sampling and lab analysis

Within the Wuliangsuhai Lake area and the wetlands of Zhangye, we sampled reed in August 2011 when it was at its peak biomass. We laid out 45 plots in the water, covering a wide range of water pollution and salinisation levels (Figure 1), and 12 plots in terrestrial habitats. The plots had an area of 0.5×0.5 m^2. Mean plant height (m) and basal diameter (cm) related to the sediment surface were measured and the stem numbers were counted. Two to five reed individuals (ramets) were sampled on each plot. Roots and rhizomes were sampled 30 cm below the sediment surface. We divided each individual into five parts (organs), i.e. flowers, leaves, stems, roots and rhizomes. The five organs were weighed on site, directly after sampling.

Additionally, water and sediment were sampled at root depths near the experimentation areas at the same time as sample collection. Using this method, for the comparison of nutrients in reed organs, we evaluated differing developmental conditions by sampling ten plots with five aquatic and five terrestrial habitats in November 2011 (Li et al. 2014a).

In the laboratory, we determined the DMC, total carbon (mg C g^{-1}), total nitrogen (mg N g^{-1}) and total phosphorous (mg P g^{-1}) contents of reed flowers, leaves, stems, roots and rhizomes. All samples were oven dried at 60 °C for 72 h for subsequent nutrient content analyses. C and N were measured using an elemental analyser (2400 II CHN Elemental Analyzer, Perkin-Elmer, USA) with a combustion temperature of 950 °C and a reduction temperature of 640 °C. P was measured following the molybdate stannous chloride method (He et al. 2008). A total ranging from 10–30 leaves and 0.5 g wet root (1–5 replicates) were scanned with a Canon scanner (4400F), per sample. Then, SLA, SRL, mean root diameter (R^{diam}) and mean root area of unit mass (R^{area}) were determined with WinFOLIA and WinRHIZO (Régent, Quebec, Canada).

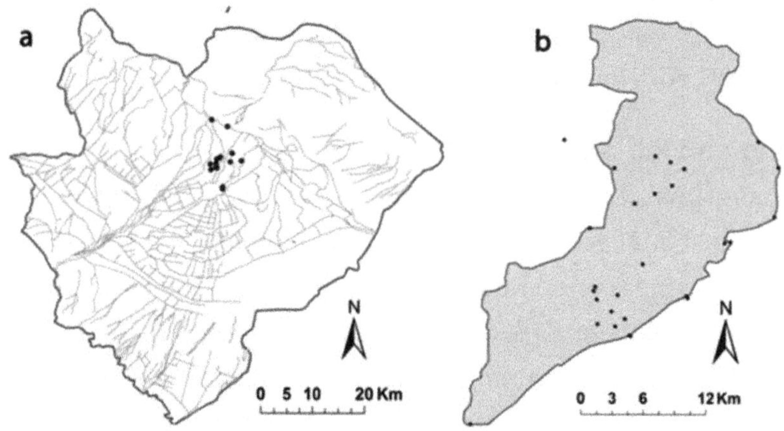

Figure 1 – Situated within northern China, locations of the reed sampling plots within the [a] wetlands in Zhangye (Zhang et al. 2005) and [b] Wuliangsuhai Lake (USGS 2011) (Li et al. 2014b).

5.2.2 Statistical data analysis

We compared the functional traits of *P. australis* from both study sites and found no significant differences. Hence, we pooled together the data from the two study sites for further analyses. One-way ANOVA and Bonferroni *post hoc* tests were carried out to assess the differences in N and P contents within and across reed organs. Standardised major axis (SMA) slopes with 95 % confidence intervals were used to examine N and P relationships within reed organs (Wright et al. 2004). Ordinary least square (OLS) regressions were performed to test the relationships of plant nutrient contents and water and sediment characteristics. ANCOVA was used to test the differences between regression slopes (Townend 2002). Since the sampling sites of the terrestrial and aquatic habitat were in close vicinity, we expected no other variations in regard to environmental conditions. The paired t-tests assess the differences in the traits of *P. australis* as well as characteristic differences between the two habitats were performed. One-sided paired t-tests were used for DMC, SLA, SRL, R^{area} and R^{diam}. DMC and R^{diam} were expected to be higher in the terrestrial reed and SLA, SRL and R^{area} higher in aquatic. Two-sided paired t-tests were used to test for differences in biomass allocation in the five *P. australis* organs and RS. In this case, the proportion of root to the total biomass was also defined as Root Ratio (RR). SMA analysis was used to detect relationships between biomass allocations in different organs. Slopes of SMA were compared between the two reed ecotypes. We considered more the trait differences of *P. australis* due to the two types of habitats (mostly moisture differences) and ignored the differences of other environmental variables, for example, soil and sediment nutrient contents. For further information on the methodological procedure, see (Li et al. 2014a, 2014b). The t-tests were used for the comparison of reed N and P content between two seasons.

5.3 Findings: N and P content

5.3.1 N and P content of reed

The N and P content of reed was on average 14.1 and 0.95 mg g^{-1}, respectively. The average N:P ratio was about 16, with 13 for the belowground and 17 for aboveground part. It differed significantly for the N and P contents in the five organs of reed. The N and P contents (mg g^{-1}) and the N:P ratio were 22.9, 2.05, and 12 in the flowers, 31.4, 1.33 and 24 in the leaves, 6.4, 0.65 and 10 in the stems, 16.1, 1.03 and 17 in the roots, and 7.2, 0.66 and 10 in the rhizomes, respectively (Figure 2). The N content was highest in the leaves while the P content was largest in the flowers compared to the other reed organs ($P < 0.05$).

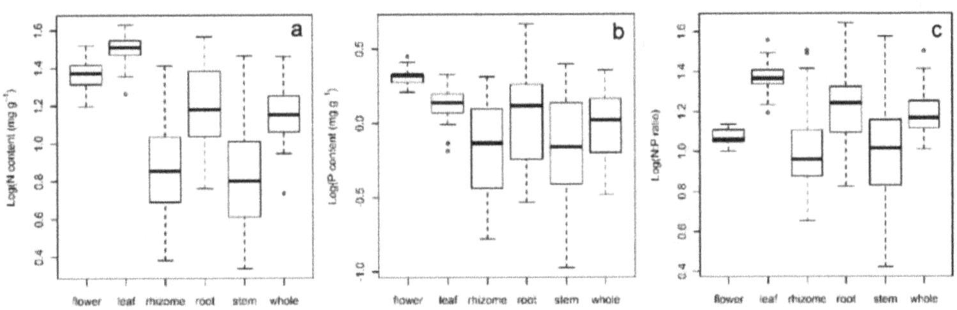

Figure 2 – Boxplot of N and P content and N:P ratio in five organs of reed with data log_{10}-transformed. The lines in the middle of the boxes indicate median values, the upper and lower ranges of the boxes show the third and the first quartiles whereas upper and lower lines out of the boxes indicate the maximum and minimum values, extreme values are shown as dots (Li et al. 2014b).

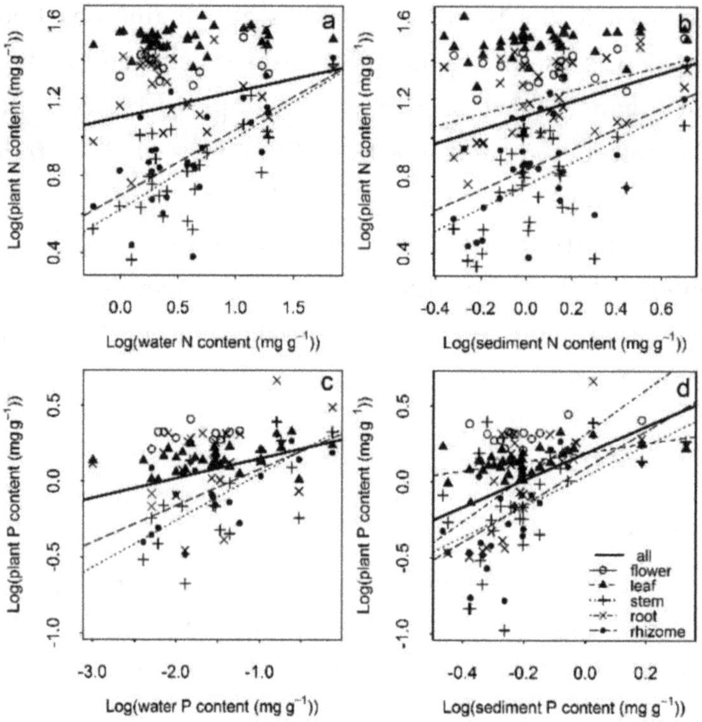

Figure 3 – The influence of water and sediment N and P availability on the N and P content of reed with data $\log_{10}$- transformed and with regression line shown when the slope is significant (P < 0.05) and not shown when the slope is not significant ($P > 0.05$) (Li et al. 2014b).

5.3.2 N and P stoichiometry of reed along environmental gradients

At the whole plant level, plant N and P contents were significantly and positively correlated with water and sediment N and P contents ($r^2 = 0.38$ and 0.18 for reed-water and reed-sediment N, respectively; $r^2 = 0.62$ and 0.43 for reed-water and reed-sediment P, respectively; all $P < 0.05$, Figure 3). At the organ level, stem and rhizome N contents increased significantly with the water and sediment N contents ($P < 0.05$, Figure 3). Flower and leaf N contents did not exhibit significant changes with the water or sediment N content ($P > 0.05$).

Stem and rhizome P contents increased significantly with the water and sediment P contents ($P < 0.05$). Root and leaf P contents increased significantly with the sediment P content ($P < 0.05$), with root P increasing fastest (slope = 1.45, $r^2 = 0.39$, $P < 0.05$) and leaf P slowest (slope = 0.31, r^2 =

0.26, $P < 0.05$). Flower P content did not exhibit significant changes with water or sediment P contents ($P > 0.05$).

5.3.3 The influence of nutrients on reed biomass allocation and community structures

Reed community structure is influenced by the environment and plant nutrient contents. Reed stem density was negatively correlated with reed P content and sediment P availability ($r^2 = 0.54, 0.17$, $P < 0.05$, Figure 4). Reed growth height and basal diameter were negatively correlated with stem density ($r^2 = 0.48, 0.67$, $P < 0.05$, Figure 4).

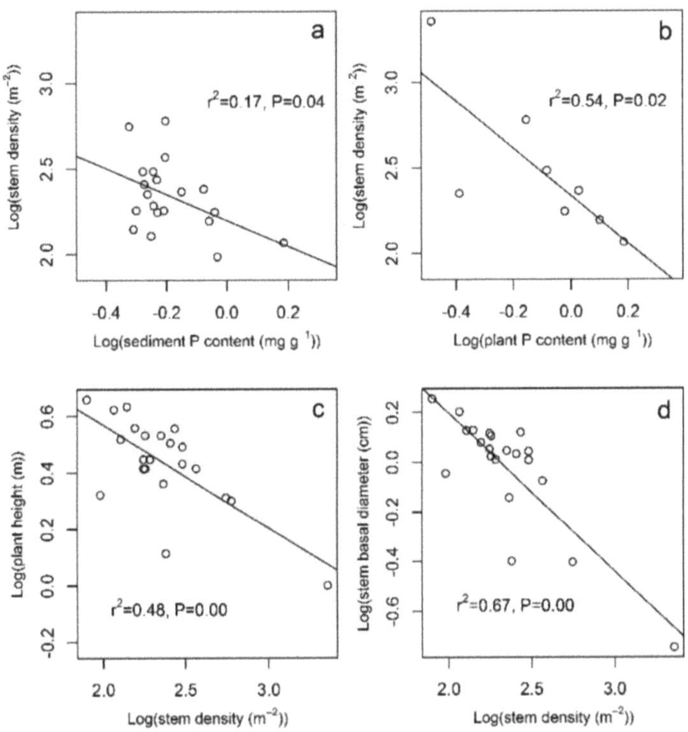

Figure 4 – The relationships of reed stem density with sediment per plant P content and reed growth height and basal diameter with stem density (with data $\log_{10}$- transformed) (Li et al. 2014b).

5.3.4 The functional trait and biomass allocation comparison of reeds in aquatic and terrestrial habitats

The mean SLA was 14.9 and 11.7 $m^2 kg^{-1}$ and the mean SRL for aquatic and terrestrial reeds were 94.9 and 40.4 m g^{-1}, respectively. The SLA and SRL were both higher for aquatic than for terrestrial reeds ($P < 0.05$, Figure 5a, b). The mean root area of unit mass (R^{area}) was greater but the mean root diameter (R^{diam}) was smaller for aquatic than for terrestrial reeds (R^{area} = 0.09 and 0.05 $m^2 g^{-1}$; and R^{diam} = 0.32 and 0.42 mm for aquatic and terrestrial reeds, respectively, $P < 0.05$, Figure 5c, d). A further illustrative comparison of reed growing in terrestrial and aquatic habitats (i.e. habitat, leaf and root) can be found in Li et al. (2014a).

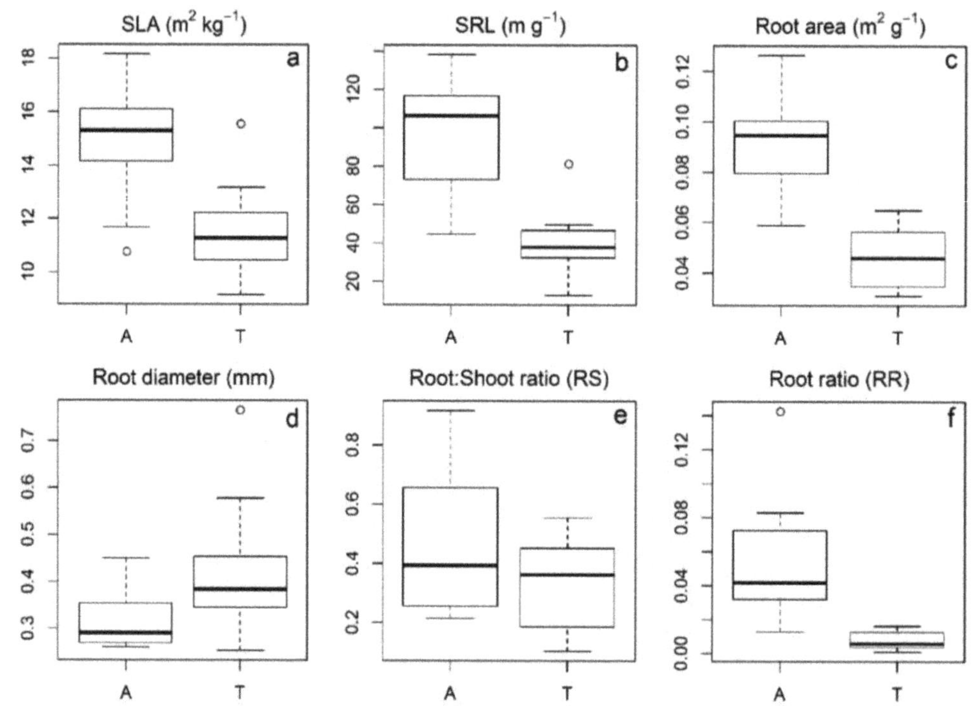

Figure 5 – The functional traits and biomass allocation of aquatic (A) and terrestrial (T) reeds with [a] = specific leaf area (SLA), [b] = specific root length (SRL), [c] = mean root area of unit mass (R^{area}), [d] = mean root diameter (R^{diam}), [e] = Root:Shoot ratio (RS) and [f] = root ratio (RR) (Li et al. 2014a).

The biomass proportions of the aboveground organs flower, leaf, and stem as well as belowground rhizome did not significantly differ for the aquatic and terrestrial plants (70 % and 76 % for aboveground biomass proportions, 3.9 % and 8.9 % for flower, 29.5 % and 36.2 % for leaf, 39.5 % and 32.3 % for stem, 21.4 % and 21.9 % for rhizome, for aquatic and terrestrial reeds, respectively, all $P > 0.05$, Table 1). The mean RS did not differ significantly between aquatic and terrestrial reeds (0.45 and 0.32, $P > 0.05$, as shown in both Table 1 and Figure 5e). However, the biomass proportion of root was significantly higher for aquatic than for terrestrial reeds (5.5 % and 0.7 %, respectively, $P < 0.05$, see Table 1 and the RR in Figure 5f).

Table 1 – The biomass allocation in reed of the aquatic and terrestrial ecotypes. Leaf:Stem indicates the biomass ratio of leaf and stem, analogously for the Root:Rhizome, Root:Leaf and Stem:Rhizome. The different superscripts indicate significant differences in the means (paired t-test, $P < 0.05$), SD = standard deviation. * The proportion of root biomass is also defined as root ratio (RR). (Li et al. 2014a).

	Aquatic form		Terrestrial form	
	Mean	SD	Mean	SD
Flower	3.9^{a}	2.9	8.9^{a}	5.9
Leaf	29.5^{a}	9.5	36.2^{a}	8.1
Stem	39.5^{a}	9.6	3.5	7.0
Root*	5.5^{a}	3.5	0.7^{b}	0.6
Rhizome	21.4^{a}	7.0	21.9^{a}	8.0
RS	0.45^{a}	0.24	0.33^{a}	0.15
Leaf:Stem	0.83^{a}	0.47	1.28^{a}	0.79
Root:Rhizome	0.26^{a}	0.13	0.04^{b}	0.02
Root:Leaf	0.22^{a}	0.18	0.02^{b}	0.02
Stem:Rhizome	2.09^{a}	0.95	1.84^{a}	1.22

Aquatic and terrestrial reeds, established in contrasting environments, have similar biomass allocation patterns but distinct leaf and root functional traits, which suggests different resource acquisition strategies. Aquatic reed grows faster with high SLA and SRL and is more receptive to the environment, while terrestrial reed is able to resist more to adverse environment and it is less responsive to the environment.

5.4 Comparative analysis of reed N and P content between summer and winter

The nutrient contents in reed organs were different between the two seasons studied (Photograph 6 and 7; Table 2). The N contents of reed's aboveground part (including leaf, stem and flower) in winter were lower than the contents in summer when reed has its peak biomass levels ($P < 0.05$). We did not find significant differences for the belowground elements (i.e. root and rhizome ($P > 0.05$)). The P contents of reed in leaf and stem in winter were lower than the contents in summer ($P < 0.05$). The P contents of reed in flower, root and rhizome were not significantly different ($P > 0.05$). The N:P ratio in stem and flower of reed was significantly different in the two seasons ($P < 0.05$). The results showed the movement of nutrient contents from leaf and stem to other organs from summer to winter.

With 10 samples (five for each of the two seasons studied), we compared the N and P contents from the winter for aquatic and terrestrial ecotypes reeds and found no significant differences in each organ. Future research and more sampling are needed to verify this result.

Table 2 – Comparison of N and P content (mg g^{-1}) of five organs of reed (aquatic form) in northern China in summer and winter. The different superscripts indicate significant differences in the means (t-test, $P < 0.05$) (Li et al. 2014a, 2014b).

	Summer N	Winter N	Summer P	Winter P	Summer N:P	Winter N:P
Leaf	31.37^a	8.01^b	1.33^A	0.34^B	$23.5^\wedge$	$23.4^\wedge$
Stem	6.37^a	1.91^b	0.65^A	0.09^B	$10.0^\wedge$	$22.0^\vee$
Flower	22.94^a	13.46^b	2.05^A	2.38^A	$11.7^\wedge$	$5.7^\vee$
Rhizome	7.18^a	6.38^a	0.66^A	0.8^A	$10.4^\wedge$	$8.0^\wedge$
Root	16.13^a	19.66^a	1.03^A	1.82^A	$17.2^\wedge$	$10.8^\wedge$

5.5 Conclusion

This research comprehensively sampled reed plants growing in different water bodies and on land in two seasons of northern China. It analysed the N and P content and biomass allocation patterns in each organ. The the N and P content of reed organs is closely related but the correlations may slightly alter with a change in environment-nutrient availability. In particular, N and P content of reed in leaf and flower is higher and more flexible, while it is lower and relatively stable in the stem. Significant increase of N and P content for aboveground parts of reed were found with an increase of sediment N and P availability, but any such corresponding increase of aboveground biomass ratio was not significant (Li et al. 2014b). Additionally, plant nutrient stoichiometry and nutrient availability are important drivers of these stand structures. The stem density of reed decreased, while stem height and basal diameter increased with the increase in nutrient availability. These results showed that with the eutrophication of wetlands and lakes, reed plants could absorb more N and P, and any extra N and P will result in more leaf and flowering of plants.

Moreover, N and P content in stem and leaf of reed decreased significantly during the winter season. This shows that, to restore a wetland ecosystem, harvesting reed in summer, when reed has its peak biomass, is not only removing more biomass but also more N and P. Limited by the sample size, we did not find significant changes of N and P content in reed belowground part for summer or winter.

Reed grows quite different in various environments, most notably is its functional trait differences in aquatic and terrestrial habitats. We found that both SLA and SRL, characteristics of leaf and root, respectively, are significantly different for reed in the two habitats even when the two grow about 10 meters in distance from each other. Being different, with the results of functional traits, we did not find very different biomass allocation patterns of reed between the two ecotypes. We preliminary concluded that this species adapted to a changing environment probably by way of eutrophication, climatic pressures and salinisation – mostly due to changes of functional traits rather than biomass allocation patterns. The scope of this research could be interlaced with climate change modelling or for the evaluation of N and P removal via reed harvesting in different eutrophication statuses for planning and restoring wetlands.

Key references

Han, W.X., Fang, J.Y., Reich, P.B., Ian Woodward, F. & Wang, Z.H. 2011. Biogeography and variability of eleven mineral elements in plant leaves across gradients of climate, soil and plant functional type in China. Ecology letters, 14(8): 788–796.

He, J.-S., Wang, L., Flynn, D.F.B., Wang, X., Ma, W. & Fang, J. 2008. Leaf nitrogen: Phosphorus stoichiometry across Chinese grassland biomes. Oecologia, 155(2): 301–310.

Kerkhoff, A.J., Fagan, W.F., Elser, J.J. & Enquist, B.J. 2006. Phylogenetic and growth form variation in the scaling of nitrogen and phosphorus in the seed plants. The American Naturalist, 168(4): 103–122.

Li, L., Han, W., Thevs, N., Jia, X., Ji, C., Jin, D., He, P., Schmitt, A.O., Cirella, G.T. & Zerbe, S. 2014a. A Comparison of the Functional Traits of Common Reed (*Phragmites australis*) in Northern China: Aquatic vs. Terrestrial Ecotypes. PloS ONE, 9(2): e89063. doi:10.1371/journal.pone.0089063

Li, L., Zerbe, S., Han, W., Thevs, N., Li, W., He, P., Schmitt, A.O., Liu, Y. & Ji, C. 2014b. Nitrogen and phosphorus stoichiometry of common reed (*Phragmites australis*) and its relationship to nutrient availability in northern China. Aquatic Botany, 112: 84–90.

Reich, P.B. & Oleksyn, J. 2004. Global patterns of plant leaf N and P in relation to temperature and latitude. Proceedings of the National Academy of Sciences of the United States of America, 101(30): 11001–11006.

Shi, B., Ma, J., Wang, K., Gong, J., Zhang, C. & Liu, W. 2010. Effects of atmospheric elevated temperature on the growth, reproduction and biomass allocation of reclamation *Phragmites australis* in East Beach of

Chongming Island. Resources and Environment in the Yangtze Basin, 19(4): 383–388.

USGS. 2011. Map of Wuliangsuhai Lake: Border of Wuliangsuhai Lake drawn from TM image acquired August 2011. [Online]. 2011. United States Geological Survey. Available from: www.usgs.gov [Accessed: 12 November 2011].

Zhang, H., Shen, W.S., Wang, Y.S. & Zou, C.X. 2005. Study on grassland grazing capacity in the Heihe River Basin. Journal of Natural Resources, 20: 514–521.

6. Diversity and role of rhizobacteria associated to reed stands (*Phragmites australis*)

Lorenzo Brusetti
Luigimaria Borruso

6.1 Introduction

Rhizobacteria are microorganisms intimately associated to root systems. They are mainly involved in plant growth promotion, supply of nutrients, defence against phytopathogens and insect biocontrol. A particular importance is recognised with the biodiversity of rhizobacteria associated to plants involved in phytoremediation. These bacteria can play an important role in pollutants and xenobiotic transformation, detoxification and mineral-isation. In this chapter, we review the biodiversity of these bacteria, under-lining their role in the phytoremediation processes, specifically focusing on *Phragmites australis*. Case studies regarding the study sites in China are discussed.

The rhizosphere of submerged plants is defined as the narrow area of sediments characterised by peculiar chemical-physical properties, directly affected by root secretions (root exudates) released by living plants. Plant exudates include amino acids, carbohydrates, sugars, vitamins, mucilage and proteins (Bais et al. 2004), responsible for chemotactical attraction of microorganisms to the plants. Nutrient availability stimulates bacterial proliferation on roots' surface (rhizoplane) and in the surrounding area (Figure 1) (Hale et al. 1971). Microbial population density in the rhizosphere can be 1,000–2,000 times higher than in the surrounding bare soil, while metabolic activity of some taxa can be incremented to more than 1,000-fold. This phenomenon is known as the 'rhizosphere effect' (Egamberdieva et al. 2008; Berendsen et al. 2012).

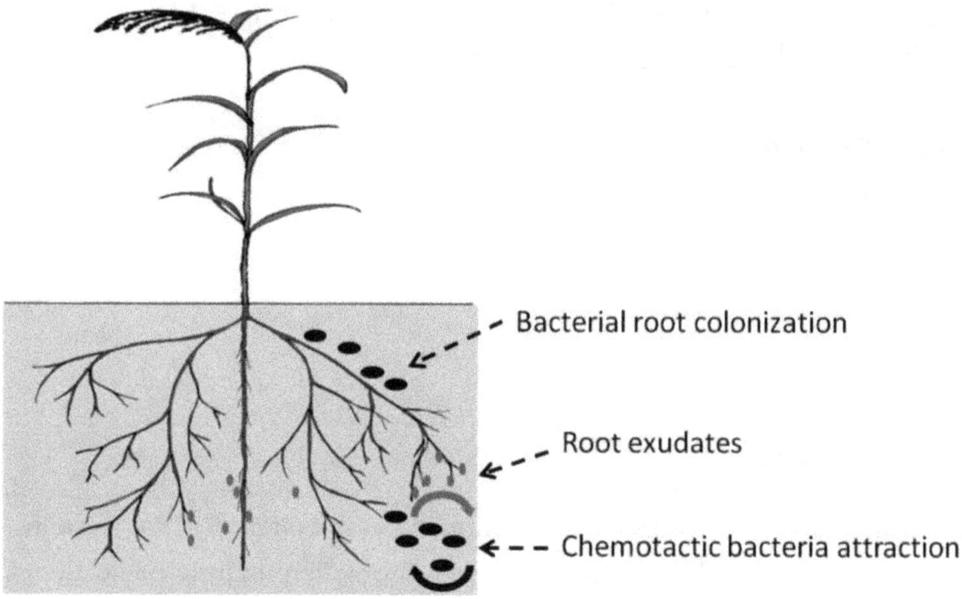

Figure 1 – Bacterial communities associated with plant roots and bacterial attraction via root exudates (Borruso 2014).

It has been reported that plant root exudation differs among species. Consequently, several experiments have found that microbial communities differ according to plant species. For example, a species-specificity of rhizobacterial communities associated to *Solanum tuberosum, Fragaria ananassa* and *Brassica napus* were found in an agricultural field (Smalla et al. 2001).

Other works reported that species-specificity of microbial communities due to the rhizosphere effect may be masked or partially masked by a variety of environmental matrices (Boeuf-Tremblay et al. 2005). In this case, particular environments such as hypersaline ponds (Lozupone & Knight 2007), where salt concentration is a stronger stressor and major responsible in shaping of rhizobacterial communities of *P. australis,* even if their root played an important role in the increment of microbial community diversity (Borruso 2014). Even other important abiotic parameters, such as soil or sediment texture (Schutter et al. 2001) or pH (Lauber et al. 2008) can sometimes play a more important role in shaping rhizobacterial communities than roots' exudates.

6.2 Microorganisms beneficial for plants

In the rhizosphere, there are a number of pathways linked to biogeochemical cycles. The exchange taking place between plants and microorganisms within the soil-root interface leads to several decomposition and mineralisation processes. Among them, the most important are symbiotic nitrogen fixation, phosphorous solubilisation, production of phytohormones, siderophores and biocontrol agents. These processes cause an increment of nitrogen and phosphorous assimilation at the root level, and a stimulation of the plant growth due to phytohormones (Glick 2012; Prashar et al. 2013).

6.2.1 Biological N_2 fixation

The most studied microorganisms involved in nitrogen fixation belong to the rhizobia group, nitrogen-fixing bacteria in symbiosis with plants of the family *Papilionaceae*. Rhizobia are a paraphyletic group ranging from α-Proteobacteria (i.e. *Sinorhizobium, Rhizobium* and *Azorhizobium*) to β-Proteobacteria (i.e. *Burkholderia, Herbaspirillum*) (Cocking 2003; Valverde et al. 2003; Udvardi & Poole 2013). Other species belonging to the genus *Azospirillum* are in symbiosis with the family of the Graminaceae, whereas *Frankia* sp. has an actinorhizal symbiosis with a nodular rhizosphere and rhizoplan of geni, e.g. *Alnus, Myricae, Betula* or *Coriaria* (Benson & Silvester 1993; Kennedy et al. 1997; Simonet et al. 1999).

6.2.2 Nutrient acquisition

Despite the high amount of phosphorous found in soil and sediments, most of it is not available (i.e. bound to calcium, aluminium or iron) for plant nutrition and growth. In this context, plants benefit of phosphate-solubilising bacteria such as *Enterobacter agglomerans* and *Azotobacter chroococcum*, which are able to convert phosphorous from an inorganic mineral form to a bioavailable form for plants (Kole et al. 1988; Kim et al. 1998). *Actinomycetae, Bacillus* and *Clostridium* play a key role in the humic acids formation by degrading lignin and cellulose plant tissues, therefore increasing soil fertility and bacterial activity (Rösch et al. 2002). In the rhizosphere, fungi, plants and bacteria compete for iron. Some bacteria, such as *Serratia, Pseudomonas, Rhodococcus* and *Acinetobacter*, are able to produce siderophores, molecules with a high affinity with ferric ions that cause iron solubilisation therefore improving its bioavailability (Prashar et al. 2013).

6.2.3 Biocontrol of plant diseases

The potential use of rhizobacteria able to stimulate plant growth and to protect plants from diseases and stresses has become a well-studied issue in the last decades. Bacteria involved in biocontrol are associated with a number of plant species widespread in many environments. These microorganisms are able to compete for space and resources, invading the pathogens ecological niche. Some strains belonging to *Pseudomonas, Bacillus* and *Stenotrophomonas* are involved in the production of antibiotics that affects phytopathogen growth and diffusion (Compant et al. 2005). For instance, *Pseudomonas fluorescens* and *P. chlororaphis* produce phenazine, responsible for the fungal disease suppression in plants (Kim et al. 2011).

6.2.4 Phytohormones production

Phytohormones are crucial in plant growth, development and response to environmental factors. Their production is a strategy adopted by bacteria to improve the interaction with roots. The production of gibberellic acid and cytokinins stimulates an increment of the dimension of root surface and length, which determines plant development positively (Kloepper et al. 2007).

6.3 Bioremediation and phytoremediation as potential tool for freshwater and sediment restoration

Freshwater sediments are considered a hot spot for the accumulation of a wide range of organic and inorganic contaminants. Metallic contaminants and nutrients represent a dramatic issue in several freshwater sediments, where the growing use of fertilisers and pesticides in agriculture has led to an increasing presence of nitrogen and phosphorous. The development of metallurgical industries has caused an increment in the level of heavy metals pollution in the environment (Su et al. 1994; Wang et al. 2001; Cheng 2003; Borruso 2014). Heavy metals negatively influence the quality of crops, atmosphere and water, or affect human and animal health (Zhang 1999; Liao

1993). Moreover, the significant presence of carbon and energy sources in sediments tends to determine an increase in biomass and diversity of microbes potentially able to degrade organic pollutants (Perelo 2010). Xenobiotics can be immobilised in sediments or transformed and degraded through biological processes carried out by microorganisms. In particular, bioremediation is the use of the metabolic potential of organisms, primarily microorganisms, to prevent or remove pollutants. There are two fundamental approaches that can be used in bioremediation, i.e. (1) the biostimulation of indigenous microbial communities through the addition of a suitable electron donor/acceptor or nutrients (Morgan & Watkinson 1989; Margesin & Schinner 2001), (2) the bioaugmentation, which is the use of an exogenous microbial community isolated from other environments and capable to break down or render harmless various contaminants (Vogel 1996).

Bioremediation can be applied only where environmental conditions permit proliferation and activity of microbial communities on site. The application often involves the variation of chemical-physical parameters to both optimise microbial growth and contaminants degradation. For example, some species of microorganisms are able to reduce metals toxicity through a number of metabolic pathways. Metals can be extracted from sediments thanks to organic acids produced by microorganisms, which can cause metals leaching by extracting them from ores into solution. *Geobacter* sp. and *Thiobacillus* sp. have the capability to transform mercury into methylated-mercury generating metal-leaching and sulphuric acid from the oxidation of elemental sulphur (Lovley 1995; Gomez & Bosecker 1999). *Penicillium simplicissimum* has the capability to transform the insoluble form ZnO into the soluble form Zn (Franz et al. 1991).

In the literature, it is reported that the role of bacteria in the bioremediation of toxic organic compounds includes polychlorinated biphenyls compounds (PCBs) and polycyclic aromatic hydrocarbons (PAHs). PCBs dechlorination is predominantly carried out by reductive processes operated by Cloroflexi. The dechlorination process has been found in different sediments affected by

PCBs contamination (Abramowicz 1995; Sowers & May 2013) and it causes a reduction in carcinogenicity potential of the PCBs (Abramowicz 1995). For example, anaerobic bacteria belonging to *Dehalococcoides* were isolated from river sediments and resulted to dechlorinate 64 PCB congeners (Bedard et al. 2007). The aromatic ring that characterises PAHs, has been recently demonstrated to be hydrogenated in anoxic conditions by bacteria that use nitrate and sulphate as terminal electron acceptors (Dou et al. 2009). *Pseudomonas* sp. and *Vibrio* sp. isolated from sediments were found to be able to degrade PAHs by using naphthalene as growth substrate (Rockne et al. 2000).

Efficiency of bioremediation may be increased by the interaction between bacteria and plant roots. Corn, wheat, macrofitae, soybean and common grasses are extensively used for bioremediation purposes due to their extensive fibrous root system extending into the rhizosphere. Usually, plants exposed to contaminants show a change in morphological and physiological traits. For instance, the growth reduction is mostly due to the increment in stress ethylene, which inhibits root elongation, nodulation and auxin transport. This effect can be mitigated via the presence of plant-grow-promoting bacteria that limit the inhibitory effects of various stresses (Glick 2010). Indeed, microbial communities strictly associated to the plant rhizosphere play a key role. Microbes play an important role in degradation, immobilisation and reduction of contaminants bioavailability (Glick 2010). It is well known that bacteria in the plant root zone break down pollutants such as organic toxins, chlorinated molecules, fuels and solvents (Mothes et al. 2010; Glick 2010).

There are different kinds of phytoremediation (Figure 2); they include:

- transportation and concentration of the substances from the environment into plant tissues (phytoextraction);
- degradation or breakdown of organic contaminants in metabolic processes carried out by plants and their associated microorganisms (phytodegradation);
- removal and absorption of toxins or excess of nutrients by plant roots from waters (rhizofiltration);
- reduction in the mobility and bioavailability of contaminants by plant roots and their associated bacteria in soil and/or groundwater (phytostabilisation); and
- plants uptake and transpiration of pollutants from soil or water, which are then released into atmosphere (phytovolatilisation) (Salt et al. 1995; Pilon-Smits 2005; Glick 2010).

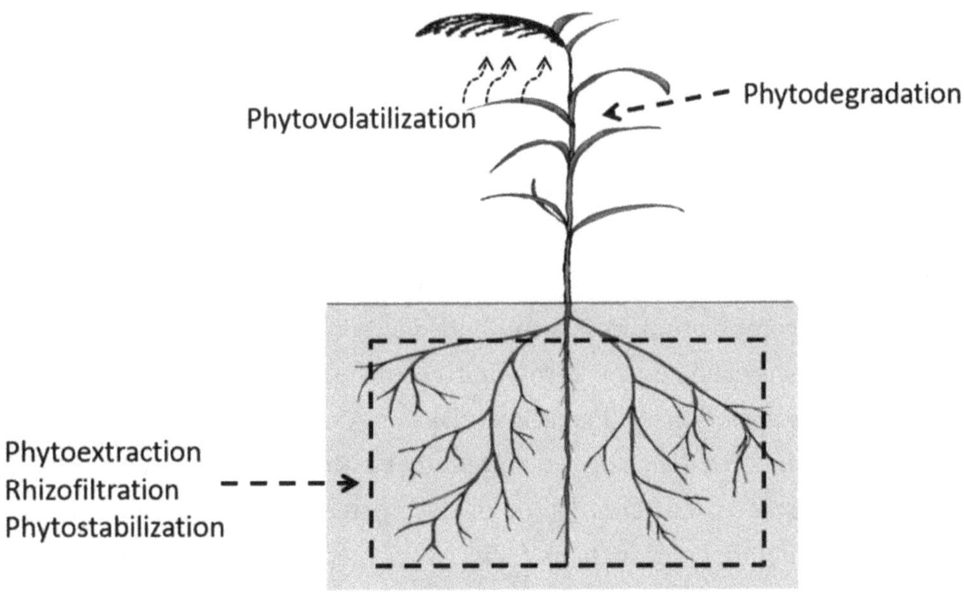

Figure 2 – Different types of phytoremediation processes involving removal and containment of contaminants (Borruso 2014).

For instance, *P. australis* stands are commonly used for phytoremediation, as they are able to collect and store in its tissues a number of toxic compounds as well as heavy metals (Southichak et al. 2006; Vymazal et al. 2009). *P.*

australis is indeed widespread in extremely polluted environments since it contains large amounts of lignin and cellulose, which are known to be able to absorb many heavy-metal ions from aqueous solutions (Vymazal et al. 2009).

6.4 Diversity and role in phytoremediation of the microorganisms associated with reed

The diversity of microorganisms associated with the rhizosphere is enormous, more than tens of thousands of species. Among bacteria, the most represented taxa are α/β/γ-Proteobacteria, Actinobacteria, Acidobacteria, CFB group and Firmicutes. Several studies regarding bacterial diversity associated with the rhizosphere of *P. australis* in natural and contaminated environments and its implications in the remediation process are reported in the literature. Sulphate-reducing rhizobacterial communities inhabiting Lake Valenci in Hungary highlighted a dominant presence of taxa affiliated to *Desulfovibrio* sp. and *Desulfotomaculum* sp. Moreover, the authors noticed a higher abundance of sulphate-reducing bacteria in the rhizosphere than in the surrounding bulk sediments probably due to the partly presence of oxygen in the rhizosphere and the root exudates (Vladár et al. 2008). In the same lake, culturable bacteria from the inner and outer rhizome surfaces were studied to assess the differences in the microbial composition between the healthy and the declining *P. australis* reed stands. A shift in the bacterial communities' composition in healthy and declining reed stands was found. Bacteria characterised by a fermentative metabolism, such as *Erwinia billingiae, Aeromonas sobria, Pantoea agglomerans, Pseudomonas azotoformans, Kocuria rosea* and *Bacillus* spp., resulted to be predominant in healthy *P. australis* reed stands. Differently, in declining reed stands bacteria with a saprotrophic metabolism such as *Acinetobacter* spp., *Aeromonas hydrophila, Curtobacterium luteum* and *Agrobacterium vitis* were isolated (Micsinai et al. 2003).

In a wetland located in northern China, Zhang et al. (2013b) analysed the rhizobacterial communities in different sites and noticed a shift according to the different levels of wetland degradation with a reduction in the abundance of Acidobacteria, Cyanobacteria and Fusobacteria and an increment in Actinobacteria in the more degraded sites. Several studies cover the role of *P. australis* and its rhizobacteria involved in phytoremediation. For example, the role of endophytic bacteria associated to *P. australis* in the Beijing Cuihu Wetland, China was analysed. The cloning library revealed that the predominant class was Proteobacteria with *Pleomorphomonas*, *Azospirillum*, and *Aeromonas*. These genera are involved in the phytoremediation of nitrogen, phosphorus, sulphur and some other organic compounds found in the wetland systems. The authors demonstrated a considerable reduction of P (56 %) total N (48 %) and organic matter (13 %) due to the presence of *P. australis* (Li et al. 2010b). Explorative results, from the samples collected in rhizosphere sediments associated to *P. australis* along the main drainage channel of the Hetao Irrigation District (see Chapter 2 and 10 for additional details), showed that the most abundant taxa were *Thiobacillus*, Nitrosomonadaceae and *Desulfobacterium*. These taxa are usually used for their bioremediation potential in particular for nitrate removal of sewage waste, of hydrocarbons degradation and on sites contaminated with chlorate such as pulp and paper industry wastewater (Borruso 2014).

P. australis has been widely used to treat wastewater from industries, medium-sized municipalities and farms (De Maeseneer 1997; Leonard & Swanson 2001; Vymazal et al. 2009) to improve water quality in freshwater environments. The municipal sewage of seven cities in the Czech Republic was treated trough filtration beds of constructed wetlands composed by *P. australis*. After the treatment, metal concentration (cadmium, nickel, lead, copper, chromium and zinc) resulted to have been decreased and become similar in concentration of unpolluted wetlands (Vymazal et al. 2010). In another study, bacterial communities in a constructed wetland located in Shanghai Mengqing Park were analysed and correlated to the different nutrients levels. A reduction of about 50 % of the ammonia and BOD_5 was reported. Moreover, a higher abundance of ammonia nitrifying bacteria in the ammonia-rich sediments revealed the key role of the microorganisms in ammonia removal (Park 2009). Additional studies showed that *P. australis* could contribute to antibiotics remediation (Conkle et al. 2008; Park et al.

2009; Hijosa-Valsero et al. 2011). Conventional wastewater and a constructed wetland treatment plants were compared to test the efficiency of antibiotic removals. After the treatment procedure, concentrations of tetracycline and Trimethoprim were evident. Decreased levels resulted in the constructed wetland treatment plant rather than in the conventional wastewater treatment plant, with a most probable cause and effect via the adsorption processes of *P. australis*. The authors explained these results with the presence of rhizobacterial communities and explained the important role and usage of antibiotic degradation (Hijosa-Valsero et al. 2011).

6.5 Conclusion

In the last few decades there has been an increasing interest by public and private institutions to manage the dangerous presence of pollutants. The traditional methods such as sediment removal, sediment oxidation or sediment filling are expensive and cause a high level of disturbance to the environment. In this context, selected plants and rhizobacterial strains due to their natural capabilities to bioremediate polluted environments can play a key role in land restoration processes. Indeed, microbial bioremediation coupled with phytoremediation processes in *P. australis* is demonstrated to be highly feasible and a well-established technology for cleaning up metals, pesticides, solvents and xenobiotics. The effectiveness of the phytoremediation approach can be increased with a better understanding of physiological, biochemical, molecular and genetic bases of the microbial communities associated to the plants used for the remediation process.

Nevertheless, a prudent policy in local territory administration that attend to *P. australis* stands both in extensive zones (e.g. lakes, large-sized ponds or marshes) and in narrower environments (e.g. natural or artificial water channels or small streams) will contribute to detoxification and remediation of highly polluted environments.

Key references

Berendsen, R.L., Pieterse, C.M.J. & Bakker, P.A.H.M. 2012. The rhizosphere microbiome and plant health. Trends in Plant Science, 17(8): 478–486.

Borruso, L. 2014. Rhizobacterial communities as bioindicators of environmental stresses in freshwater ecosystems. PhD Thesis. Bolzano, Italy: Free University of Bozen-Bolzano.

Cheng, S. 2003. Heavy metal pollution in China: Origin, pattern and control. Environmental science and pollution research international, 10(3): 192–198.

Glick, B.R. 2012. Plant Growth-Promoting Bacteria: Mechanisms and Applications. Scientifica, 2012: 1–15.

Glick, B.R. 2010. Using soil bacteria to facilitate phytoremediation. Biotechnology Advances, 28(3): 367–374.

Hale, M.G., Foy, C.L. & Shay, F.J. 1971. Factors affecting root exudation. Advances in Agronomy, 23: 89–109.

Hijosa-Valsero, M., Fink, G., Schlüsener, M.P., Sidrach-Cardona, R., Martín-Villacorta, J., Ternes, T. & Bécares, E. 2011. Removal of antibiotics from urban wastewater by constructed wetland optimization. Chemosphere, 83(5): 713–719.

Pilon-Smits, E. 2005. Phytoremediation. Annual review of plant biology, 56: 15–39.

Simonet, P., Navarro, E., Rouvier, C., Reddell, P., Zimpfer, J., Dommergues, Y., Bardin, R., Combarro, P., Hamelin, J., Domenach, A.M., Gourbière, F., Prin, Y., Dawson, J.O. & Normand, P. 1999. Co-evolution between *Frankia* populations and host plants in the family Casuarinaceae and consequent patterns of global dispersal. Environmental Microbiology, 1(6): 525–533.

Smalla, K., Wieland, G., Buchner, A., Zock, A., Parzy, J., Kaiser, S., Roskot, N., Heuer, H. & Berg, G. 2001. Bulk and rhizosphere soil bacterial communities studied by denaturing gradient gel electrophoresis: plant-dependent enrichment and seasonal shifts revealed. Applied and Environmental Microbiology, 67(10): 4742–4751.

7. Reeds as a renewable energy source: Insight into possible conversion pathways

Marco Baratieri
Francesco Patuzzi

7.1 Introduction

This chapter deals with the possible use of reed for energy production, with particular regard to innovative thermochemical conversion processes as pyrolysis, gasification and – among upgrading pathways – torrefaction. After introducing the role of wetland biomasses, these energy conversion processes are described in the first section, explaining also their possible role as alternatives to direct combustion of biomass. Different scenarios of the complete reed-to-energy chain are here presented. Beside physical and chemical characteristics of a feedstock, a crucial issue to be considered in assessing the reed suitability for energy production is the feedstock thermal behaviour. For this purpose, a detailed thermal analysis of reed has been performed through combined analytical techniques and the results are presented in the second section of this chapter. The third section is devoted to the investigation of different conversion scenarios of reed, through modelling analysis supported by experimental tests of torrefaction and pyrolysis carried out in a laboratory bench-scale apparatus. The results described in this chapter allow a general assessment of reed as energy source, and they have been used for the techno-economic evaluation of a case study in Wuliangsuhai Lake region, presented in Chapter 8.

7.2 Biomass as a source of energy and the role of wetland biomasses

In the last years, the interest on renewable energy sources has been on the rise. On the one side, the energy demand is constantly growing, in strict connection to the global population rising and the expansion of developing countries' economies (Nelson 2011). This aspect, combined with the depletion of fossil fuels, caused in the last few years a considerable increasing on the global energy markets of the prices for fossil fuel energy. On the other side, industrialised societies are more and more aware of impacts of fossil fuel utilisation on the environment and on the human health, making the search for environmental and socially acceptable alternatives increasingly important (Kaltschmitt et al. 2007). Respect to other renewable sources such as wind or solar energy, biomass has the main advantage that, if well managed, can ensure a constant supply of energy, being its availability not dependent on meteorological conditions. This is an essential aspect in the vision of an integrated exploitation of different renewable sources.

Beside woody biomass, perennial grasses have gained widespread appeal for energy production, opening new renewable energy scenarios (Strezov et al. 2008). Perennial grasses are now being used as solid fuel in co-fired coal power plants as well as targeted as a choice feedstock for such advanced biofuels as cellulosic ethanol. Furthermore, perennial grasses can also be pressed into pellets, briquettes, and cubes and used as a heating fuel to replace or complement fuels made from wood fibre. Grasses can grow on marginal lands suited for continuous row crop production or in open rural land currently not in agricultural production. They usually have a high productivity and once established, require far fewer inputs in comparison to annual crops in terms of fertiliser and pesticides needs.

In addition to terrestrial grass plants, also the emergent aquatic macrophytes can play an important role as local energy source in places where they are naturally present in great amounts. Wetland plants have a high macro-nutrients accumulation capability because of their generally fast growth and high biomass production (Bragato et al. 2006; Frick et al. 2011). As a consequence, their utilisation for energy production could have, as additional advantage, the disposal of such kind of elements from the ecosystem. In the present research, these considerations have been in particular applied to the emergent aquatic macrophyte *Phragmites australis*, known as common reed. Common reed is one of the most common plants living in wet ecosystems (Bonanno & Lo Giudice 2010; Dienst et al. 2004; Ostendorp 1999). Among other grass plants, common reed is considered to be a promising source of renewable energy, being a perennial fast-growing plant able to produce an interesting amount of dry biomass (3–15 t/ha) and being its provision relatively cheap (Lee 2011).

P. australis can withstand extreme environmental conditions, including the presence of toxic contaminants such as heavy metals (Baldantoni et al. 2004; Quan et al. 2007; Ye et al. 1997; Bonanno & Lo Giudice 2010) and there are many cases in which, as other wetland plants, it is utilised for the removal of pollutants, including metals. Concerning the phytoremediation of metals, there are generally two possible approaches, i.e. (1) "phytostabilization", where the plants are used to immobilise metals and store them belowground in roots and/or soil, in contrast to (2) "phytoextraction" in which hyper-accumulators may be used to remove metals from the soil and concentrate them in aboveground tissues (see Chapter 6). These latter plants must be, in turn, harvested and disposed of to prevent recycling of accumulated metals when the plants decompose (Weis & Weis 2004). Metal translocation into the shoots appears to be very restricted in *P. australis* so that harvesting plants will not be an effective source of metal removal in a wetland system.

From an ecotoxicological perspective, it could also be argued that the transfer of metals into shoot biomass is an undesirable property, as metals so accumulated could pass into the food chain via herbivores and detritus feeders (Ye et al. 1997). However, *P. australis* show a high N and P concentration in tissues and may have greater potential to remove nutrients from eutrophicated wetlands. Phosphate is easily concentrated in the belowground tissues, while nitrate concentration is higher in leaf and stalk. Harvesting the aboveground tissues could take most of the nitrate out of the sediment (Tian et al. 2009).

Focusing on common reed, in the international literature few experiences can be found concerning its utilisation for energy production. One of the first studies on reed utilisation for energy purposes dates back to the 1980s in Sweden, in particular for the application of reeds in direct combustion (Graneli 1984). Nevertheless, in the following years, even if this species has been widely studied for its phytodepuration potential (Weis & Weis 2004; Bonanno & Lo Giudice 2010; Baldantoni et al. 2004; Quan et al. 2007; Ye et al. 1997), few experiences can be found in international literature concerning the energy utilisation of *P. australis*. In fact, the low bulk density (20–60 kg m^{-3}) of reed makes it unattractive for long distant transportation and increases handling and storage costs.

Only in the last years, the interest on this topic rose again, mainly in Europe (Kuhlman et al. 2013). Most investigations deal with the combustion of reed (Barz et al. 2006; Kask et al. 2007; Kitzler et al. 2012; Komulainen et al. 2008; Kronbergs & Kronbergs 2011), some with biogas production (Hansson & Fredriksson 2004; Komulainen et al. 2008). Besides this, a promising but not yet completely characterised application of common reeds seems its thermochemical conversion through pyrolysis (Sutcu 2008; Zhao et al. 2011; Park et al. 2012) or gasification (Kitzler et al. 2011; Yang et al. 2012; Link et al. 2013), in the perspective of converting the biomass into liquid or gaseous fuels that can be used in a more efficient way respect to the direct combustion of a solid fuel. Sutcu, in 2007, carried out some experiments about pyrolysis of common reeds in a tubular fixed-bed system, achieving the maximum oil yield at 550 °C pyrolysis temperature, a heating rate of 25 °C/min and a sweeping gas flow rate of 100 ml/min (Sutcu 2008). The feedstock ultimate and proximate analyses are in good agreement with the result of Lange (2006) and the comments followed by Barz et al. (2006). They indicate three points: (1) compared to other biomass fuels, the relative high LHV indicates the reed as a promising energy source; (2) the nitrogen content is very low so that no problems concerning nitrogen oxide emissions were expected; (3) the higher contents of chloride, sulphur and ash might cause problems regarding emissions and process management if the reed is used in conventional combustion technologies.

7.3 Thermochemical conversion of biomass

Biomass can be converted into energy by means of different methods, as shown in Figure 1. Focusing only on thermochemical conversion processes, the easiest application is represented by direct combustion in a furnace, after simple mechanical pre-treatments (e.g., chopping, pressing, briquetting). The released heat can be then used for steam generation and energy (heat and electricity) production. Nevertheless, traditional combustion of solid heterogeneous fuels is affected by several environmental impacts such as the release of particulate matter and unburned hydrocarbons that can be somehow mitigated through the adoption of expensive clean-up stages downstream the process.

As an alternative, pyrolysis and gasification processes can be used in order to convert the biomass into gaseous (i.e. producer gas) or liquid fuels that can be used in a more efficient way. For example, cleaner combustion technologies (i.e. premixed burners) can be used, as well as high-efficiency internal combustion engines (e.g. alternative engines, gas turbines).

7.3.1 Processes: Torrefaction, pyrolysis and gasification

Many of the problems in biomass gasification are related to the properties of the fuel. Biomass is thermally unstable and has a low energy density and, as a consequence, its gasification leads to tar formation and low energy products. For this reason, in the last few years biomass torrefaction process has gained a renewed attention on the biomass-to-energy production chain. This thermal pre-treatment is a sort of slow pyrolysis process carried out at low temperature (max. 280–300 °C) that makes better physical and thermochemical properties of biomass (Bergman & Kiel 2005; Arias et al. 2008; Bridgeman et al. 2008), such as grindability, uniformity, hydrophobicity and heating value – and can be a suitable pre-treatment to improve the gasification/pyrolysis of the feedstock (Prins et al. 2006).

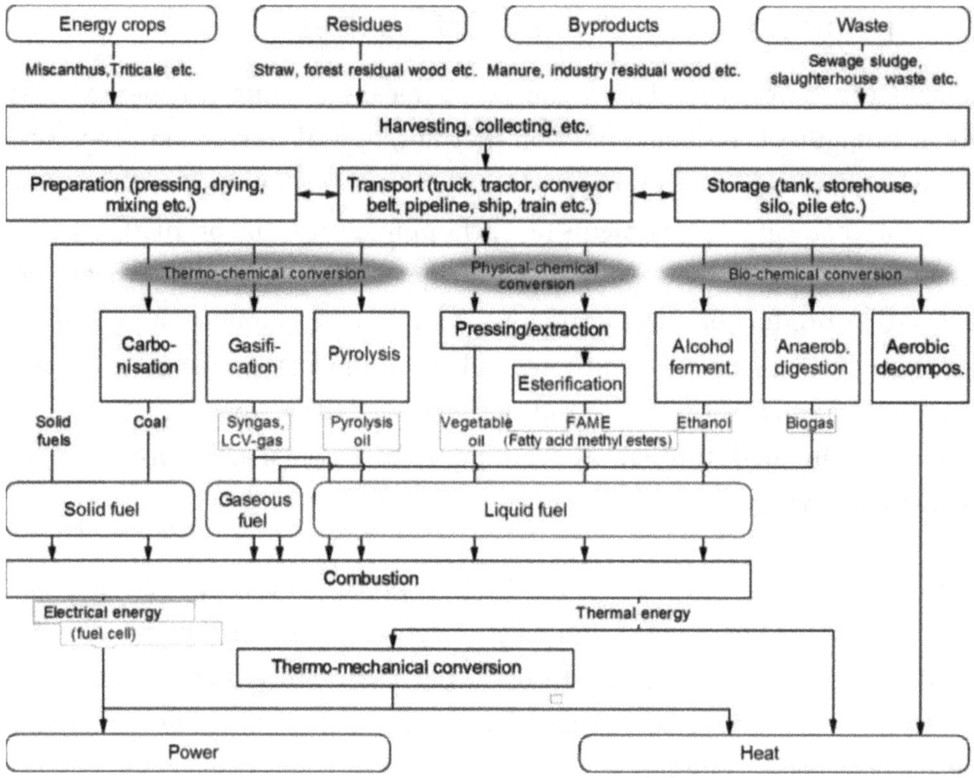

Figure 1 – Schematic representation of different biomass-to-energy pathways (rounded boxes: energy carriers; boxes: conversion processes); adapted from Kaltschmitt et al. (2007).

Pyrolysis is a thermal treatment carried out in absence of oxygen that allows to transform the biomass into liquid (pyrolytic oils and tar), gaseous (synthesis gas or syngas) and solid (char or biochar) products. Tar, a complex mixture of condensable hydrocarbons of high molecular mass, can cause operational problems in downstream processes by blocking gas coolers, filter elements and engine suction channels (Li & Suzuki 2009). Recently, char or biochar starts to be considered, besides as fuel, as a valuable compound useful to increase soil fertility and carbon sequestration in soil (Kwapinski et al. 2010). On the basis of the heating rate, one of the main parameters beside the peak temperature, it is possible to distinguish between "fast" and "slow" pyrolysis. The aim of the former, characterised by high heating rate (up to 1000 °C per second) and peak temperature of about 650 °C, is the maximisation of the liquid product yield. In Table 1 the typical product yields of pyrolysis carried out at different process conditions are reported.

Table 1 – Typical product yields %, (dry wood basis) obtained by different modes of pyrolysis of wood (Bridgwater 2004).

Pyrolysis conditions	Liquid	Char	Gas
Moderate temperature, short residence time	75	12	13
Low temperature, very long residence time	30	35	35
High temperature, long residence times	5	10	85

Pyrolysis could also be seen as the first step in combustion and gasification processes. In fact, if an oxidant agent is introduced, the pyrolysis is followed, respectively, by the total or by the partial oxidation of the primary products (Figure 2). The sub-stoichiometric conditions that characterise the gasification process can be reached through the injection of steam (in the case of "steam gasification"), air or oxygen (in the last two cases the process is generically named "partial oxidation").

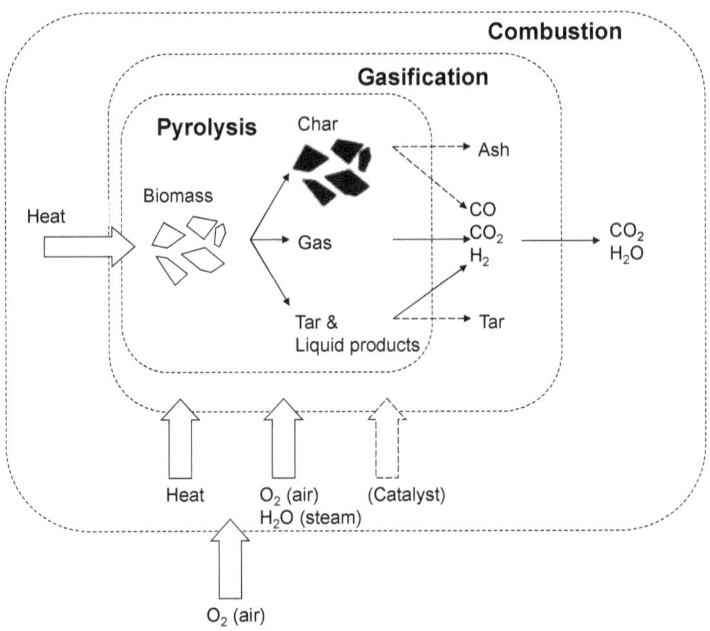

Figure 2 – Schematic representation of pyrolysis as the first step of gasification and combustion; readapted from Knoef (2005).

7.3.2 Overview on biomass conversion technologies in China

Nowadays, China is one of the countries with the most dynamic economies in the world. The energy necessary to ensure this fast economic growth is mostly produced from coal (74 % in 2011, according to the International Energy Agency), which is cheap and easily available. The drawback is the high emission of greenhouse gases. Producing "clean" energy is one of the most challenging tasks for China in the 21st Century.

China tries to tackle this challenges by increasing the share of renewable energies from 8 % in 2013 to 15 % in 2020 (Moch 2013). Whereas wind, photovoltaic and hydropower has been mostly used for electricity production, biomass is often burned in rural areas for heat production (Moch 2013). Latter role is comparable small, only 1 % of the total energy consumption in 2002 (Zhang 2012), and up to now relies mostly on small-scale technology, e.g. stoves for local heat production by combustion or digestion. Nonetheless, biomass resources (agricultural residues, grassland biomass and forest residues) can have a relevant role on substituting coal, being their energy potential equal to almost 30 % of China's energy consumptions (Zhou et al. 2011). According to the China 12th Five-Year Plan for Renewable Energy, the production of biogas in rural areas is intended to increase to 50 million households by 2015 (Moch 2013). Seven billion tonnes (t) of solid waste are produced annually, with a proportion of 60 % biomass waste (EC2 2011). The available amount of crop straw corresponds to 0.82 billion t, but 20 % are burned in the field without being utilised (Zhang 2012).

Up to now in China, in the rural areas, household energy is mainly produced by means of direct coal or biomass combustion (Zhang et al. 2013a) with methods characterised by very low energy efficiencies, e.g. 10 % for the traditional stoves burning crop residues and 7 % for the Kang, a kind of coupled stove and heated bed system in rural northern China (Junfeng et al. 2005; Liu et al. 2008), as shown in Photograph 8a, b. This causes serious indoor air pollution related to respiratory aerosols, SO_2 (in particular when coal is burned) and CO emission (Zhang & Smith 2007).

Because of the high amount of biomass feedstock in China, several research institutes and universities are devoted to research, development, demonstration and application on biomass gasifier (Zhang et al. 2013a). According to the review of technological development of biomass gasification for a variety of applications in China performed by Leung et al. (2004), biomass gasification for the purpose of electrical energy generation is very promising, possessing great potential in research and development in China. The main biomass gasification demonstration projects are summarised in Table 2.

Table 2 – Biomass gasification demonstration projects in China (Zhang et al. 2013a).

Type	Feedstock	Purpose	Capacity	Location
Downdraft	Sawdust	Drying	200 kWt	Beijing
	Crop residues	Cooking	300 kWt	Shangdong
	Wood residues	Cooking	700 kWt	Liaoning
	Crop residues	Cooking	300 kWt	Hunan
	Agricultural residues	Heat and power	200 kWe	Jilin
Bubbling Fluidised Bed	Crop residues	Cooking and heating	300 m^3/h	Henan
	Rice husk	Electricity	400 kWe	Anhui
	Rice husk	Heating	120 MW	Jiangsu
	Rice husk	Gas	160-220 m^3/h	Anhui
Circulating Fluidised Bed	Rice husk, rice and wheat stalk	Electricity	5500 kWe	Jiangsu
	Sawdust, rice husk or straw	Electricity	1000 kWe	Fujian
	Wood powder	Heating	1000 kWt	Guangdong
	Wood powder	Heating	7000 kWt	Jilin
	Crop residues	DME	1000 t/a	Guangdong
	Wood powder	Electricity	1000 kWe	Hainan
	Corn straw	Electricity	600 kWe	Hebei

Figure 3 – Location of the sampling points in Wuliangsuhai Lake (Inner Mongolia, China); digitised background from Landsat satellite images.

7.4 Characterisation of reed as energy source: chemical-physical properties and advanced thermal analysis

Several samples have been collected in Wuliangsuhai Lake wetland area (Figure 3) and subsamples from stems have been characterised by means of elemental and calorimetric analysis. The choice to focalise the analyses only on stems come from the weight distribution in *P. australis,* that – according to

Shi et al. (2010) – is more or less 47 % in roots, 37 % in stems, 15 % in leaves and 1 % in inflorescences.

The samples collected are only of aboveground tissues, therefore these percentages became 70 %, 28 % and 2 % for stems, leaves and inflorescences, respectively. This weight distribution is also confirmed by Li et al. (2013). The moisture content of the collected samples has been determined according to the UNI CEN/TS 14774-2. Stem subsamples were dried at 45 °C for 48 hours and ground with a Retsch Mill MM400. Ash content (UNI EN 14775), elemental composition (UNI EN 15104), sulphur and chlorine content (UNI EN 15289), phosphorus (UNI EN 15290) and Lower Heating Value (UNI EN 14918) have been determined.

Table 3 – Chemical and physical characterisation of *P. australis* stem tissues. IDs refer to the sampling point position as shown in Figure 3, S = storage (Köbbing et al. 2014b).

	IDs	2	3	4	5	6S	7	7S	8	9	10	11	12	13
Moisture	$\%wt^{ar}$	16.0	32.2	62.4	47.0	12.2	30.9	12.8	12.7	20.9	27.5	24.0	51.7	20.0
Ash	$\%wt^{dry}$	3.8	4.2	3.3	4.4	4.3	4.5	2.7	7.1	7.0	7.8	3.3	6.8	5.2
LHV	MJ/kg^{dry}	17.25	16.90	17.52	16.85	17.17	17.43	17.82	16.76	16.96	16.73	17.38	16.44	17.03
C	$\%wt^{daf}$	49.3	49.3	48.8	47.8	49.1	49.4	49.0	49.0	48.7	48.7	49.2	49.2	49.3
H	$\%wt^{daf}$	6.4	6.5	6.4	6.4	6.5	6.4	6.4	6.5	6.2	6.4	6.6	6.7	6.4
N	$\%wt^{daf}$	0.3	0.4	0.3	0.3	0.5	0.5	0.3	0.3	0.3	0.4	0.2	0.4	0.5
O	$\%wt^{daf}$	43.0	42.5	43.4	44.5	42.9	43.0	44.0	43.4	43.9	43.3	43.4	42.4	42.6
S	$\%wt^{daf}$	0.1	0.1	0.2	0.3	0.2	0.1	0.1	0.1	0.1	0.1	0.1	0.2	0.1
Cl	$\%wt^{daf}$	0.9	1.2	0.9	0.7	0.8	0.6	0.2	0.7	0.8	1.1	0.5	1.1	1.1
P	mg/kg^{dry}	119	287	61	103	116	176	82	181	86	126	71	240	175

In addition, in Table 4 the metals' content, determined according to the UNI EN 15297, of the three selected samples is reported. In fact, this is another important feature to be considered, since reeds are able to accumulate pollutants due to their phytodepuration potential (Weis & Weis 2004) and after thermochemical conversion processes most of the metals will be concentrated in the ashes.

The thermal behaviour of reeds has been characterised using a simultaneous thermogravimetric analyser (STA 449F3 Netzsch). This technique combines both the heat flux differential scanning calorimetry (DSC) and thermogravimetry (TG). STA analyses were performed at a constant heating rate of 20 °C/min under inert nitrogen atmosphere in a temperature range from 30 to 1000 °C. The evolved gas was further characterised by TG/STA coupled to a gaschromatography-mass spectrometry system and a Fourier Transform Infrared Spectrometer (FT-IR). TG and DTG curves of stems and leaves are shown in Figure 4.

Table 4 – Metals' content and ash melting behaviour of *P. australis* stem tissues. IDs refer to the sampling point position as shown in Figure 3, S = storage (Köbbing et al. 2014b).

IDs	3	4	6S
metals [mg/kg]dry			
As	0.8	0.4	0.7
Cd	0.2	0.1	< 0.1
Cr	5.0	1.2	3.3
Cu	1.0	0.7	0.6
Hg	< 0.1	< 0.1	< 0.1
Mn	62.5	71.2	104.4
Ni	1.5	1.3	1.1
Pb	0.7	1.0	0.5
ash characteristic temperatures [°C]			
deformation	842	839	871
hemispherical	1,047	954	978
flowing	1,115	1,092	1,094

All the profiles show three main reaction mechanisms. The first peak around 100 °C is due to water loss and can thus be attributed to the residual water content of the biomass. Water loss is followed by two reactions (B at around 280–300 °C and C at around 350–380 °C) which are also due to the main weight loss during thermal degradation. Figure 5 shows the Total Ion Current (TIC) chromatogram evaluation at the temperatures corresponding to these two degradation steps.

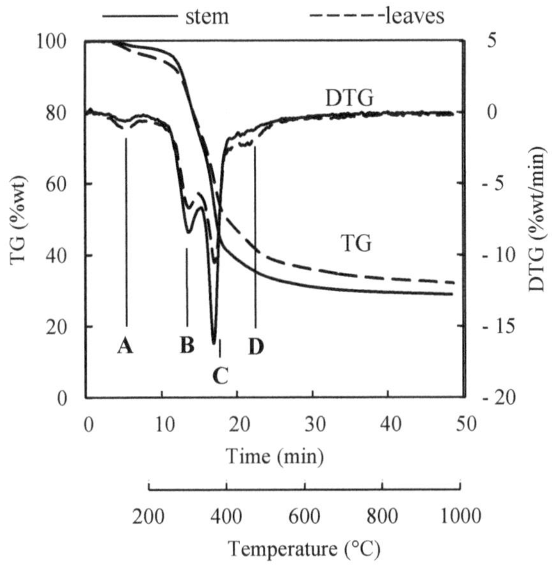

Figure 4 – Comparison between TG and DTG curves of stems and leaves of reed. Circles indicate the temperature at which IR spectra have been extracted; adapted from Patuzzi et al. (2013a).

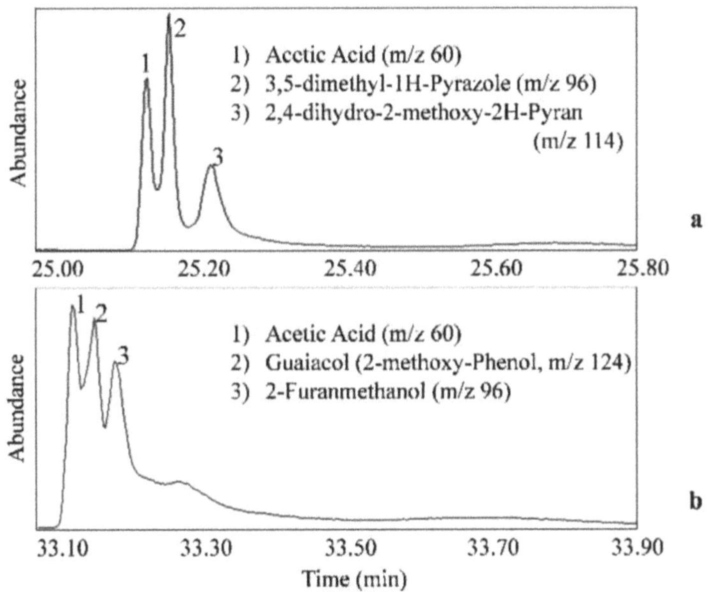

Figure 5 – Total Ion Current (TIC) chromatogram evaluation at [a] 2nd TG step (≈ 280 °C) and [b] 3rd TG step (≈ 350 °C) (Patuzzi et al. 2013a).

The residual mass at 700 °C is 30.17 % and 34.63 % for stems and leaves, respectively; as a consequence, considering the average biomass distribution in *P. australis* proposed by Shi et al. (2010), i.e. about 71 % in stem and 29 % in leaves, the weighted average conversion is 31.5 %.

In order to better define the quality of the evolved gas during thermal conversion, FTIR spectra were recorded at four selected temperatures (Table 5), corresponding to three main reactions during thermal degradation. There are no qualitative differences between the IR spectra of stem and leaves. As example, the results for the stem samples are shown in Figure 6.

Table 5 – Temperatures at which FTIR spectra of the various samples have been recorded: A moisture loss; B first reaction; C second reaction; D almost completed reactions; adapted from Patuzzi et al. (2013a).

Plant part	temperature [°C]			
	A	B	C	D
stem	102.4	281.7	353.7	429.8
leaves	101.6	280.7	352.2	556.0

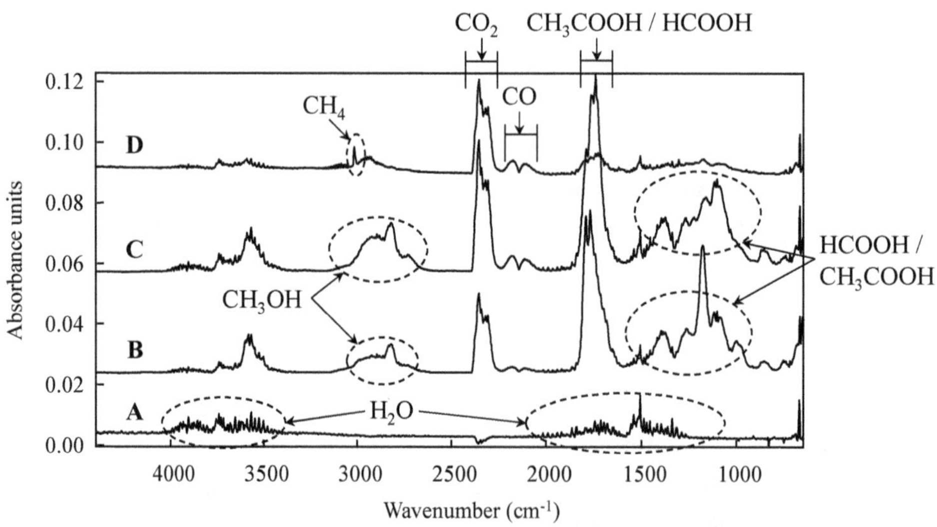

Figure 6 – FT IR spectra of the stem sample (A ≈ 105 °C, B ≈ 280 °C, C ≈ 350 °C, D ≈ 560 °C); adapted from Patuzzi et al. (2013).

The moisture loss of biomass samples attributed to the first peak (A) was confirmed by the FTIR spectra recorded at 100 °C. At higher temperatures, i.e. around 300 °C methanol (CH_3OH), acetic acid (CH_3COOH) and formic acid ($HCOOH$) could be detected. These three compounds are still present up to 350 °C but not at temperatures higher than 450–500 °C. In addition, carbon dioxide (CO_2) release increased from temperatures around 300 °C to temperatures around 350 °C and then decreased at around 450–500 °C. At this latter temperature, methane (CH_4) could also be detected in the evolved gases. This is in accordance to thermodynamic equilibrium calculation, as well as the increasing trend of carbon monoxide (CO).

7.5 Reed conversion scenarios: Experimental and modelling investigation

7.5.1 Experimental analysis

Several pyrolysis experimental tests have been performed in a bench scale apparatus setting different process temperatures (300 °C, 450 °C, 550 °C, 600 °C and 700 °C), setting the heating rate of the oven equal to 50 °C min^{-1}. The bench scale pyrolysis apparatus – set up at the Free University of Bozen-Bolzano and schematically presented in Figure 7 – consists of a tubular electric furnace, a tubular reactor, a condenser, a series of impinger bottles for the collection of tars and a module for gas sampling and measurement. The furnace is able to reach the maximum temperature of 1050 °C. The reactor (20 mm diameter, 45 mm length) is placed inside the furnace. Reactor and condenser are made out of quartz, which has a high resistance to temperature and chemicals, high purity and high stability.

The atmosphere inside the reactor is kept inert by nitrogen flow. The gas produced inside the reactor flows through the condenser, where it is cooled. The tar collection is then performed in a series of six impinger bottles, according to the technical specification UNI CEN/TS 15439; the first impinger bottle acts as a moisture collector; all the bottles except the last one are filled with isopropanol, an organic solvent suitable for tar capture. Except the first and the fourth, all the bottles are equipped with G0 frits (i.e., sintered glass filters with a nominal pore size in the range 160–250 μm). The first, second and fourth bottle are kept at 35/40 °C with water as cooling liquid, while the others are cooled at –15/–20 °C with a mixture of salt/ice/water. The gas suction device consists of a drying tower, a rotameter, a dry running rotary vane vacuum pump and a dry gas volume meter (measurement range: 0.4–6.0 m^3h^{-1}; accuracy: 0.09% at 1.2 m^3h^{-1}, –0.2% at 0.4 and 6.0 m^3). For each test, the tar–isopropanol solution has been collected and analysed by GC–MS technique. PAH have been separated on a DB5 MS column (J&W Scientific) and detected by a high resolution (R > 10.000) mass spectrometer (GC–HRMS, MAT95XL, Thermo Scientific). PAH have been identified by the addition of deuterated internal standards.

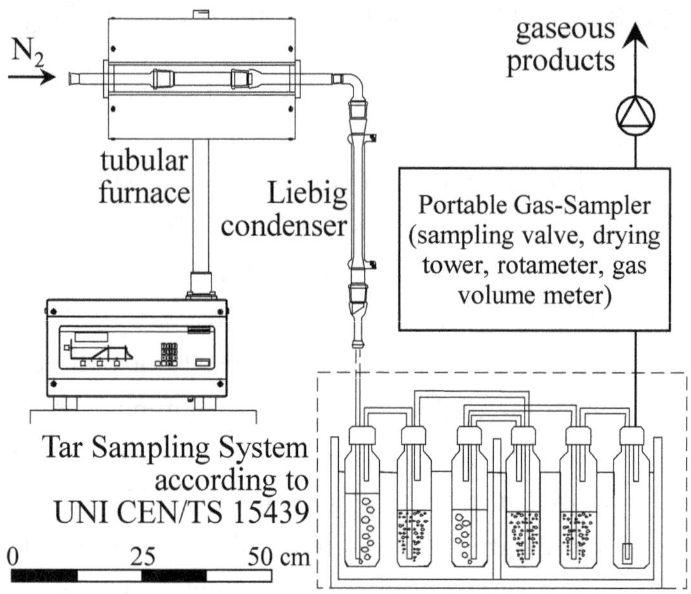

Figure 7 – Set-up of the bench-scale batch pyrolysis/torrefaction system (Patuzzi et al. 2013b).

The detected PAH concentrations are reported in Figure 8. The main detected compounds are: naphthalene, phenanthrene, fluoranthene, acenapthene, acenaphthylene, dibenzo[al]pyrene, indeno[cd]pyrene and pyrene. The total PAH concentration, as well as the concentration of most of the main compounds, show a clear increasing trend with the temperature. This behaviour is in good agreement with the literature (Garcia-Perez & Metcalf 2008).

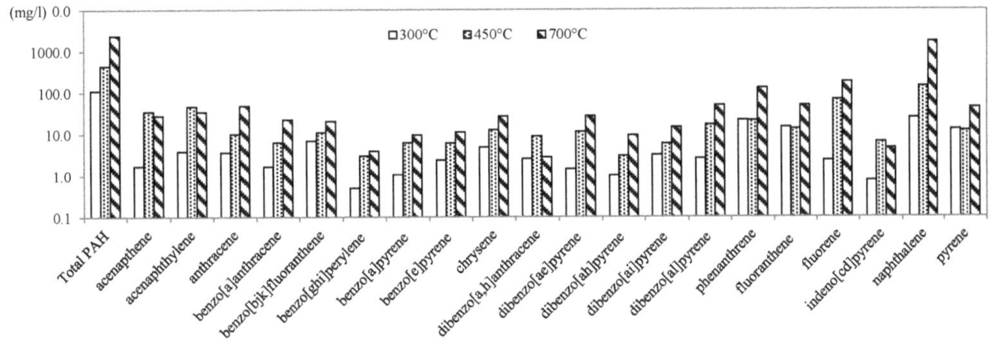

Figure 8 – PAH concentration in the condensed tar samples (Patuzzi et al. 2013b).

The total PAH content has been normalised respect to the initial biomass sample, obtaining an increasing PAH yield with the temperature (2.75 mg g^{-1}, 15.84 mg g^{-1} and 74.41 mg g^{-1} at 300 °C, 450 °C and 700 °C, respectively). In pyrolysis tar, a PAH content ranging from 7.0 to 12.6 % at 400 °C to 21. 4 % at 800 °C have been reported for lignocellulosics (Ku & Mun 2006).

7.5.2 Modelling of a CHP plant based on reed gasification

Starting from the composition and the lower heating value of the reed samples collected both in water and on the land, the average composition and lower heating value have been calculated (Table 6), assuming a moisture content of 10 %, a typical value that can be reached after an air-drying pre-treatment. The average elemental composition and calorific value have been used as input in multi-stage model of a CHP biomass plant (Prando et al. 2013). The power plant layout (Figure 9) consists of a first section where the producer gas is produced and a second section where the producer gas is exploited to generate electrical and thermal power. The producer gas production section has been modelled as a fixed bed air gasifier operating in ideal conditions which can simulate the pyrolysis and air gasification process. The chemical reaction can be endothermic or exothermic depending on the process conditions. In the first case, the heat is provided by a burner, which is fed through a producer gas spilling, while for exothermic operation the heat is supposed to be discharged. Before feeding the CHP plant, producer gas is piped through heat-recovery and clean-up sections.

The first stage of the model is the gasification unit, simulated through an enhanced gas-solid equilibrium approach, previously tested versus experimental available data and based on the minimisation of the Gibbs free energy (Baratieri et al. 2008). This allowed the estimation of the theoretical yield and equilibrium composition of the reaction products (producer gas and char) of reed thermochemical conversion processes (pyrolysis and gasification). The model consists of a code written in Matlab environment that handle chemical reaction equilibria, implementing the Cantera software library, a collection of object-oriented software tools for problems involving chemical kinetics, thermodynamics and transport phenomena (Goodwin 2011).

Figure 10 shows the producer gas molar composition for reed pyrolyzed at different temperatures and a process pressure of 1 bar. According to Baratieri et al. (2008), methane and carbon dioxide formation are favoured at lower temperatures, and at higher temperatures carbon monoxide and hydrogen are the dominant equilibrium products. This is mainly due to the increasing importance of the endothermic water gas reaction, which causes a corresponding decrease of the producer gas steam content (generated by the drying of the feedstock). Furthermore, carbon dioxide trend goes through a maximum according to its exothermic formation and endothermic conversion.

The equilibrium composition versus the equivalence ratio (ER, ratio between the oxygen fed to the gasifier and the stoichiometric quantity of oxygen needed for the complete oxidation of the species) has also been evaluated (Figure 11). The increase of the ER parameter causes a proportional increase of the N_2 molar fraction due to the greater amount of air fed to the reactor (Baratieri et al. 2008).

Table 6 – Average composition of reeds (as received basis), used as input of the thermodynamic equilibrium model (Köbbing et al. 2014b).

moisture	ash	C	H	N	O	S	LHV
[%wt^{ar}]							[MJ/kg^{ar}]
10.0	4.7	39.6	5.2	0.3	40.1	0.1	15.07

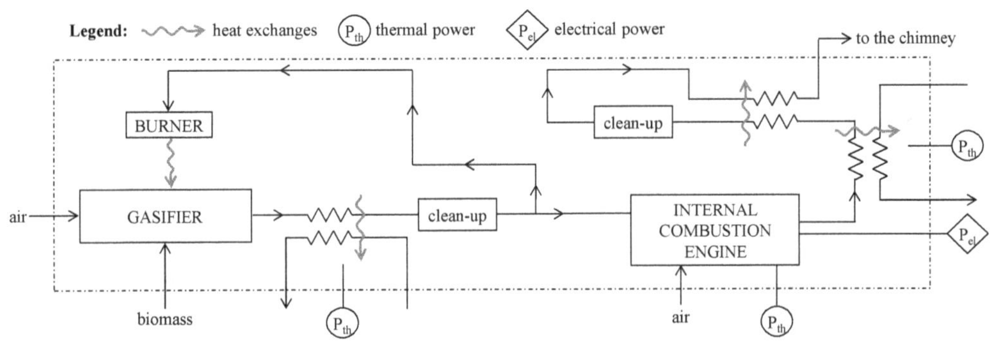

Figure 9 – Schematic representation of the considered CHP power plant layout; adapted from Prando et al. (2013).

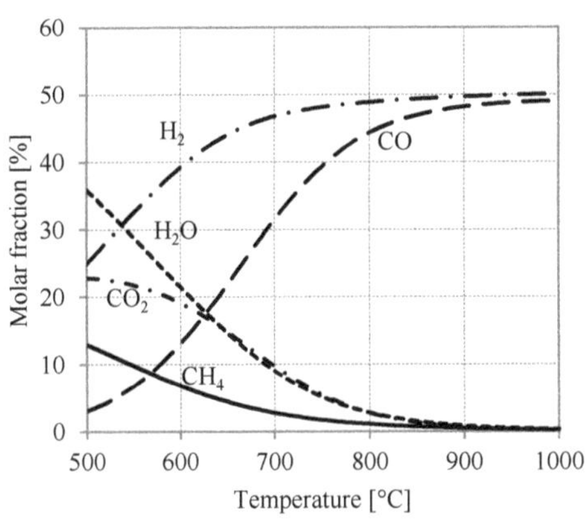

Figure 10 – Producer gas equilibrium composition versus process temperature under pyrolysis conditions and pressure of 1 bar (Köbbing et al. 2014b).

The producer gas heating value represents the useful energy output from the gasification process. An increase of the amount of gasifying agent tends to lower the producer gas LHV, mainly decreasing its CH_4 and H_2 molar concentrations. The global enthalpy variation along the conversion process has been computed taking into account all the enthalpy fluxes i.e. enthalpies of the biomass, reaction products and gasifying agents. This quantity, if positive (or negative), represents the amount of energy that has to be supplied to (or is released by) the process itself. The net process enthalpy variation is an increasing function of the process temperature. In addition, a rising of ER determine an increase of the partial oxidation process exothermicity (Baratieri et al. 2008).

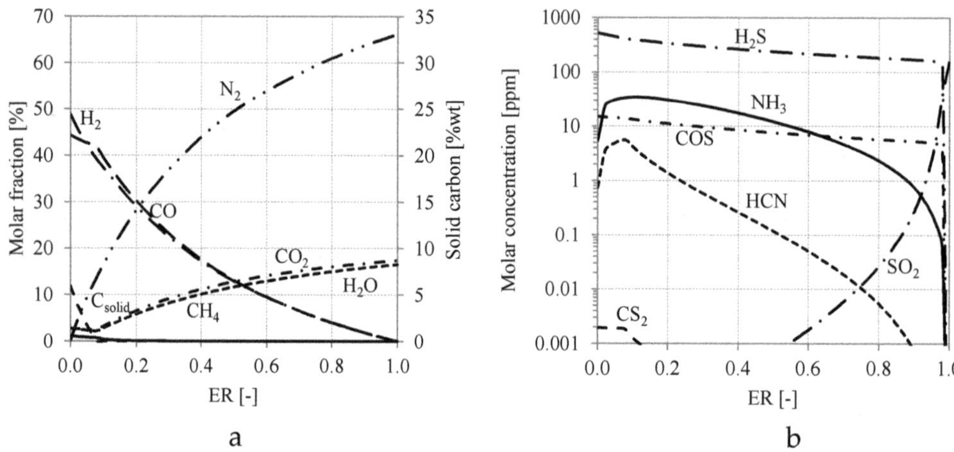

Figure 11 – Producer gas equilibrium composition and solid carbon versus equivalence ratio under partial oxidation and steam gasification of reed at 800°C and 1 bar; [a] major and [b] minor components (Köbbing et al. 2014b).

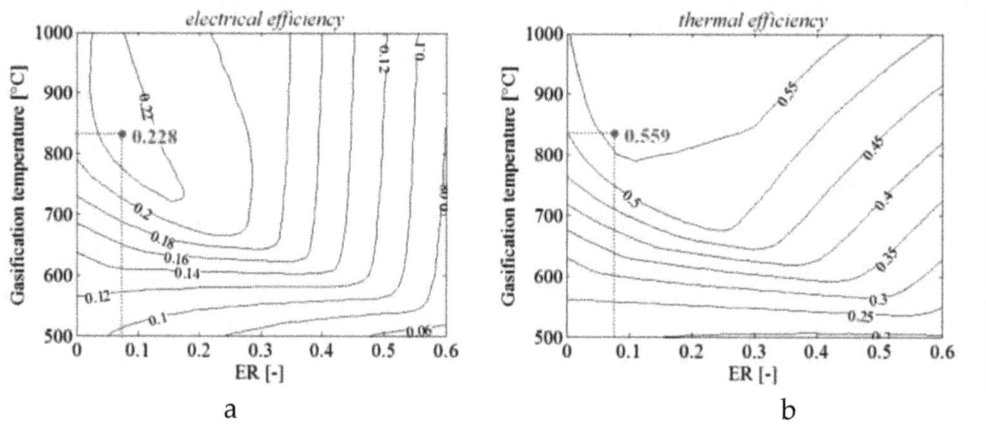

Figure 12 – Contour line representation of the CHP [a] electrical and [b] thermal efficiencies that can be achieved under different gasifier operating conditions. The highlighted values are the ones assumed for the scenario evaluation (Köbbing et al. 2014b).

The efficiency of the whole CHP system has been evaluated supplementing the simulation of the gasification stage with the energy balance of the cogeneration set (i.e., internal combustion engine). The electrical efficiency is maximised when the gasifier is operated at 870 °C and ER=0.06. Under these conditions the thermal and electrical conversion efficiency results equal to 55.9 % and 22.8 %, respectively (Figure 12). These values have been used in the scenarios evaluated in Chapter 8.

7.6 Conclusion

Several reed samples collected within the Wuliangsuhai Lake wetland area have been characterised by means of elemental and calorimetric analysis. The results show that reed is characterised by a remarkable energy content that is comparable to the one of traditional woody biomass. Nonetheless, the high ash content and the relatively low ash melting temperatures are both aspects that should be carefully taken into account in the design of a thermal degradation plant.

In addition, the application of advanced analytical techniques (TG/STA-IR-GC-MS), supplemented by the experimental pyrolysis tests at different temperatures, in a bench scale apparatus, have given useful insight into the thermal degradation process of common reed (*P. australis*). The STA analysis highlighted two main degradation steps, one at around 280 °C and the other at around 350 °C. According to the pyrolysis tests, the PAH concentration in the collected tar, as well as, the concentration of most of the main compounds shows a clear increasing trend with the temperature.

Finally, starting from the measured reed composition, the reed-to-energy pathways through pyrolysis and gasification processes have been further assessed applying a thermodynamic equilibrium approach. Based on both modelling and experimental results, *P. australis* can be considered an attractive natural biomass resource both acting as a second generation feedstock and as phytoextractant for its metal accumulation potential.

Key references

Baratieri, M., Baggio, P., Fiori, L. & Grigiante, M. 2008. Biomass as an energy source: thermodynamic constraints on the performance of the conversion process. Bioresource Technology, 99(15): 7063–7073.

Barz, M., Ahlhaus, M. & Wichtmann, W. 2006. Energetic Utilization of Common Reed for Combined Heat and Power Generation. In: 2nd International Baltic Bioenergy Conference. 2006, Stralsund, 168–175.

Bridgwater, A. V. 2004. Biomass fast pyrolysis. Thermal Science, 8(2): 21–49.

Kaltschmitt, M., Streicher, W. & Wiese, A. 2007. Renewable Energy: Technology, Economics and Environment. Berlin: Springer Berlin Heidelberg.

Köbbing, J. F., Patuzzi, F., Baratieri, M., Beckmann, V., Thevs, N., & Zerbe, S. 2014b. Economic evaluation of common reed potential for energy production: A case study in Wuliangsuhai Lake (Inner Mongolia, China). Biomass and Bioenergy, 70, 315–329. doi:10.1016/j.biombioe.2014.08.002

Knoef, H.A.M. 2005. Practical aspects of biomass gasification. In: H. A. M. Knoef ed. Handbook Biomass Gasification. Meppel: Biomass Technology Group BV.

Patuzzi, F., Mimmo, T., Cesco, S., Gasparella, A. & Baratieri, M. 2013a. Common reeds (*Phragmites australis*) as sustainable energy source: Experimental and modelling analysis of torrefaction and pyrolysis processes. GCB Bioenergy, 5(4): 367–374.

Patuzzi, F., Roveda, D., Mimmo, T., Karl, J. & Baratieri, M. 2013b. A comparison between on-line and off-line tar analysis methods applied to common reed pyrolysis. Fuel, 111: 689–695.

Prando, D., Patuzzi, F., Pernigotto, G., Gasparella, A. & Baratieri, M. 2013. Biomass CHP Systems for Residential Applications: A Multi-Stage Modeling Approach. In: K. Kabele, M. Urban, K. Suchý, & M. Lain eds. CLIMA 2013 - 11th REHVA World Congress and the 8th International Conference on Indoor Air Quality, Ventilation and Energy Conservation in Buildings. June 2013, Prague: Society of Environmental Engineering (STP), 3309–3318.

Zhang, K., Chang, J., Guan, Y., Chen, H., Yang, Y. & Jiang, J. 2013a. Lignocellulosic biomass gasification technology in China. Renewable Energy, 49: 175–184.

8. Livelihood and economy of reed wetlands

Jan Felix Köbbing
Niels Thevs
Francesco Patuzzi
Marco Baratieri

8.1 Introduction

Reed, in particular common reed (*Phragmites australis*), is an important worldwide wetlands plant covering an area of more than 10 million ha worldwide (Allirand & Gosse 1995). The biggest extensions can be found in the Scandinavian countries, Kazakhstan and China. Providing a wide range of ecosystem services, local people are in particular interested in the raw material utilisation. Reed has been used for centuries as a fodder plant, as well as a construction material for houses, gardens and boats (Köbbing et al. 2013, 2014a). In recent times, it has been used for pulp and paper production, roof thatching as well as for energy feedstock.

China with an area of 40 million ha of wetlands and a reed biomass yield of 2.6–2.7 million tonnes (t) in 2004 is the biggest reed processing country in the world. Concentrated in the northwest, north, northeast and coastal east the harvested reed is almost exclusively used for paper production. By these estimates, reed usage contributes to local livelihood and, in particular, within rural areas. Wuliangsuhai Lake, Inner Mongolia, in the north of China, is a case in point for a rural utilisation of reed. Utilised for paper production, the reed economy is put under stress from rapid rising wages and tightening environmental standards. Some recommendations and an outlook on solutions and new applications are provided, focusing in particular on energetic applications.

8.2 Reed biomass potential: Analysis worldwide and in China

Species of the genus *Phragmites* Trin. are distributed around the world, except for the Artic. The most prominent species also worldwide is *Phragmites australis* Trin. ex. Steud. As a pioneer plant it is able to spread very fast into new areas and build mono-species stands. Highly tolerant against water level fluctuation and tolerant against salinity, in can grow on sites with a groundwater level of up to 3 m below surface (Thevs et al. 2007), periodically inundated sites, and along shorelines in 1–2 m deep water.

On sites with very favourable conditions, i.e. inundated by shallow water, high temperature and high solar radiation, *P. australis* reaches a NPP of 30 t ha^{-1} a^{-1} (Köbbing et al. 2014a). Thereby, as a natural plant it does not require treatment like irrigation, seeding, weeding or herbicide or pesticide treatment. Therefore, *P. australis* offers a huge potential as a valuable resource for rural people especially in developing countries, who use it for all kind of applications, e.g. mats, baskets, roofs, fodder, pulp production or as an energy source. *P. australis* biomass also might be used as a source for polymers and other chemical products (Fachagentur für nachwachsende Rohstoffe 2012; Smole et al. 2013).

The potentially available reed biomass is lacking reliable data and is difficult to quantify. As a natural plant it is not part of official statistics such as agricultural crops and residues. Moreover, many former reed beds were reclaimed for farmland, urbanisation etc. in the last decades which makes it even difficult to rely on literature data. Nevertheless, we have attempted an initial estimation based on the available data worldwide and on China, in particular (Köbbing et al. 2013).

Table 1 summarises the available data about reed beds. Some of the sources are rather old but, if we assume that only half of the total area identified still exists, it would exceed 4 million ha. If we could harvest only half of it, yielding 5 t ha^{-1}, the quantity of biomass made available would be 10 million t annually. Atchison (1995) estimated the worldwide available reed resources to be 30 million t in 1989.

Table 1 – Reed areas and yields from winter harvest in different countries. HB = Harvested biomass [t ha^{-1} a^{-1}] (Köbbing et al. 2013).

Site, region or country	Reed bed area [ha]	HB	Total yield [t]	Year	Reference
Europe					
Poland	60,000	-	-	-	Rodewald-Rudescu (1974)
South Finland	30,000 (15,000 harvestable)	10	150,000	2006	Komulainen et al. (2008)
South Sweden	230,000	5	1,150,000	2012	Iital et al. (2012)
Mecklenburg-West Pomerania, Germany	1,500	-	-	1997	Schäfer (1999)
The Netherlands	9,000 (2,850 harvested)	-	-	-	Sluis et al. (2013)
Lake Neusiedl, Austria	60,000 (36,000 harvestable)	7	28,500	-	Schuster (1985); Gamauf (2000) cited in Kitzler et al. (2012)
United Kingdom	7,700 managed for conservation	-	-	2013	Mills (2013)
Estonia	27,899 (12,970 harvestable)	7	88,368	-	Kask (2011)
Only lakes, Latvia	13,200 (10,826 harvestable)	7.2	69,000	2009/10	Cubars (2010)
Curonian Lagoon, Lithuania	4,995	-	-	2012	
Kaliningrad Oblast, Russia	200–300	-	-	-	Iital et al. (2012)
Regions and provinces of Russia	> 1,715,000	-	-	1959	
Kazakhstan	2,000,000	-	-	1959	Krivitzki (1959) cited in Rodewald-Rudescu (1974)
Uzbekistan	800,000–1,000,000	-	-	1959	
Turkmenistan	1,000,000	-	-	1959	

Azerbaijan	50,000–100,000	-	-	1959	
Danube Delta, Romania	190,000 (125,000 harvested)	5	625,000	1965	Rudescu et al. (1965) citied in Dela Cruz (1987)
Danube Delta, Ukraine	105,055	5	50,000	-	Rodewald-Rudescu (1974)
Hungary	26,200	-	-	-	Ruttkay et al. (1964) cited in Rodewald-Rudescu (1974)
America					
Brackish, salt and tidal marshes, USA	1,800,000	-	-	1991	Chambers et al. (1999)
Asia					
NW, N, NE and coastal east China	484,000	5.5	2,600,000–2,700,000	2004	Pöyry (2006)
North & South Korea	30,000 and 20,000	-	-	-	Rodewald-Rudescu (1958)
Iraq	17,300	-	-	2000	UNEP (2001)
Globally	>10,000,000	-	-	-	Allirand & Gosse (1995)

According to Pöyry (2006), which were the last available data, *P. australis* covered an area of 1 million ha (484,000 planted reed) outside of protected areas in China 2004. Large reed bed areas are concentrated along the east coast, in the Liaohe River Basin, and in the river basins of the Yangtze and Yellow River. Further contiguous areas are distributed in Inner Mongolia and Xinjiang (Pöyry 2006). The total annual harvested biomass was 2.6-2.7 million t of reed (5.5 t ha^{-1}, moisture content 15-18 %) in 2004, headed by Hunan with 650,000 t and Liaoning with 470,000 t (Table 2).

Table 2 – Reed area and yields (incl. 15-18 % moisture content) in the provinces of China 2004 (Pöyry 2006).

Province	Yields [t ha^{-1}]	Reed area [ha]	Total yields [t a^{-1}]	Reed price after harvest [€ /t^{-1}]	Reed price at end user [€ /t^{-1}]
Heilongjang	1.3	160,000	210,000	2-3	5-6
Hunan	10.0	63,000	650,000	-	-
Liaoning	7.5	66,000	470,000	4	-
Xinjiang	9.0	43,000	400,000	2.5-3	-
Hubei	9.7	36,000	350,000	5	-
Inner Mongolia	4.0	26,000	100,000	-	3
Jiangsu	0.7	113,000	80,000	5	-
Jilin	0.2	470,000	110,000	-	-
Shangdong	7.5	13,000	100,000	3	-
Anhui	5.3	15,000	80,000	-	-
Total	5.5	1,005,000	2,550,000	3.4	4.6

The area is estimated to reach 541,000 ha in 2010, dating back to 1998 this is an increase of 1.2 % per year with an annual biomass yield of 4.2 million t of reed (Zhu et al. 1998). A productivity increase from 4 t ha^{-1} to 10 or 12 t ha^{-1} was estimated to be possible until 2010 (Pöyry 2006; Zhu et al. 1998). Table 3 shows that reed can be an important raw material in China with big concentrated stands, which allows comparable easy harvest.

8.3 Reed products worldwide: Source for products

Reed has been used for centuries as raw material for construction, fodder, fertiliser and as an energy source. Today these four reed utilisations still exist and new applications, e.g. raw material as polymers, are investigated. On an industrial scale, reed is used in the field of house construction, paper production and as an energy source (as discussed in Chapter 7). As house construction material, reed plays a role for roofing.

Table 3 – Reed area and yields in China from 1985-2010. * Forecast from 1998, ** Pöyry (2006) (Zhu et al. 1998).

Year	Area [ha]	Yields in plantation [million t]	Yields in natural reed beds [million t]	Total yields [million t]	Supply for pulp [million t]	Pulp output [million t]
1985	442,670	1.36	-	1.36	-	-
1987	471,410	1.47	-	1.47	-	-
1988	483,300	1.62	-	1.62	-	-
1989	488,680	1.65	-	1.65	-	-
1990	481,650	1.66	-	1.66	-	-
1991	485,510	1.69	-	1.69	-	-
1992	488,000	1.73	-	1.73	-	-
1995	510,000	1.95	0.61	2.56	2	-
2000*	540,000	2.62	0.65	3.27	2.78	1.08
2004**	484,000	-	2.6-2.7	-	-	1.1
2010*	600,000	4.2	0.75	4.85	4.2	-

For a long time, reed has been the only roofing material in northern Europe. Until present, it is used for roof thatching in a wide part of northern Europe and Japan. The main consumers are The Netherlands, Germany, Poland, Denmark, United Kingdom, the Baltic countries and Japan (Hawke & Jose 1996; Stenman 2008). Production has shifted from local domestic production to large reed areas in the Danube Delta and recently China. Straight, dry, nutrient poor winter reed is required and pursued for € 2 to € 3 per bundle. In house construction reed can also act as material for insulation and reed panels (Holzmann & Wangelin 2009). Panels can be used to separate walls, fixed at walls for insulation and covered with clay. Such panels can be produced in a flexible way according to the sizes required. Cost vary from around € 6.50 to € 10 /m^{-2}, e.g. as in Austria (Reichel 2013).

Reed can also be chopped and compressed to granulate panels suitable for indoor construction (Reichel 2013; Holzmann & Wangelin 2009). Next to house construction material, paper mills are the other big consumer of reed at the industrial scale. The high cellulose content (i.e. 39-59 %) of reed biomass makes it a demanding source for pulp production (Rodewald-

Rudescu 1958). Challenges occur with the sewage treatment of non-wood pulp waste water (Paavilainen 1998), which are one of the reasons for the shutdown of reed paper mills in Sweden, Egypt, Romania, Iraq, Italy, the former German Democratic Republic and the former U.S.S.R (Wayman 1973; Sainty 1985). Today, such reed mills operate in China and parts of India (Savcor Indufor Oy 2006). One ton of paper pulp requires 5.3 m^3 of softwood, 4.1 m^3 of beech wood or 3.3–3.5 t of reed (Chivu 1968). The last introduced possible industrial utilisation is the use of reed as a basis for bio-based plastic. The high cellulose content makes it a possible material for functional polymers (Fachagentur für nachwachsende Rohstoffe 2012). Lignin, hemicelluloses and especially cellulose are extracted and used for different applications, e.g. viscose/rayon, plastic or ethanol (Holzmann & Wangelin 2009); though, this application of cellulose-biomass is still in its research stages.

Reed biomass is converted into energy through combustion, biogas or as a bio-fuel. All parts of a reed plant can be used for energy generation. For combustion purposes we find the dry reed, from winter harvested plants, the preferred choice. The dry reed biomass is fired in stoves or power plants. The bulk density (20–60 kg m^{-3}) of reed is low so that a use in a local scale is favoured, in order to avoid long distance transport. Still, reed biomass is pressed into bales, briquettes or pellets, in order to increase its density (Iital et al. 2012; White 2009). The average heating value of reed is 17 MJ kg^{-1} and the calorific value of wood pellets is 17 MJ kg^{-1} whereas that of oil is 42.5 MJ kg^{-1} (BIOMASS Energy Centre 2013), which is around half that of coal (Köbbing et al. 2014b). Biogas is produced in anaerobic digestion from green spring or summer reed, which have a high nutrient content (Kask 2011). One kilogram of reed produces 0.4-0.5 m^3 of biogas with a maximum methane content of 55-60 % (Komulainen et al. 2008). Reed as raw material for biogas digestion is suitable for domestic digesters as well as large-scale plants; the left over sludge, if not polluted, can be used as fertiliser (Hansson & Fredriksson 2004). The production of biofuels (bio-ethanol or bio-diesel) is possible from all kind of biomass with a high cellulose content, which can be pre-treated to glucose (Tutt & Olt 2011). For reed, this application is still in an experimental stage due to missing demand, high costs and lacking of availability. Reed can be eaten by water buffalo, sheep, cattle and goat either as fresh fodder or as hay (Häkkinen 2007; Thevs et al. 2007; White 2009). It is easy to digest (similar to hay or straw) and can be a roughage (especially K and Mn) for ruminants, but has a comparable low nutritional value (Baran et al. 2002; Rodewald-Rudescu 1974). As a fertiliser, only pre-treated reed is suitable, for example as sludge from biogas digesters. Preferably summer reed should be used as fertiliser, because it contains sufficient amounts of nutrients (Hansson & Fredriksson 2004). Untreated, only chopped reed biomass, is unsuitable as fertiliser due to its high C:N ratio.

Next to the applications of harvested reed biomass, reed beds have the ability to purify water. This function is used in many natural as well as artificial wetlands to treat nutrient pollution from nitrogen and phosphorus. The nutrients are removed by bacterial ammonification and dentrification processes initiated by reed stalks and by nutrients uptaken by the reed plants (Kronbergs et al. 2006). If reed biomass is harvested in spring or summer, the nutrients incorporated in the biomass are removed from the specific wetland. In autumn, the nutrients are relocated into the rhizome so that not many nutrients are removed through winter harvest. Thus, reed harvest as restoration technique removes nutrients from wetlands at the same time. The nutrient peak of the aboveground biomass is reached in July and August, whereas in winter the nutrients are stored in the root system. Experience from Sweden show an extraction of 20 kg N and 1 kg P in a reed yield of 5 t ha^{-1} in winter time (Graneli 1990).

8.4 Reed usage in China: Important feedstock

The utilisation of reed has a long tradition in China since centuries. Traditional uses are fodder, mats, baskets, huts, construction material, straw checker boards for sand fixation and fire starting material (Hansmann 2008b). Currently about 95 % of the harvested reed is used for paper production, minor uses are still mats or fodder. Reed is harvested in winter, in particular in north China where an ice cover allows a simple harvest. Mostly cut by local farmers, the reed is bundled, pressed and transported to the end user, mainly paper mills (Photograph 9). Reed for fodder is grazed or cut by boat in late spring or summer (Photograph 10). Paper production, the most important application for reed in China is investigated. Approximately 2.6 to 2.7 million t of raw reed material has been used in 2004, corresponding to 1.1 t of pulp. This amount equals to 10 % of the non-wood fibre pulp production in 2004 (Figure 1). According to the estimations of Zhu et al. (1998), this would correspond to 4.2 million t of reed in 2010 (Table 3).

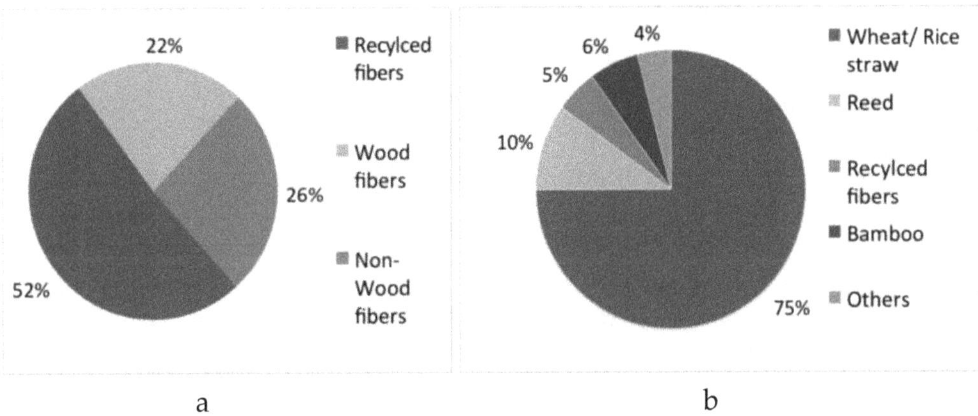

Figure 1 – [a] Total pulp production and [b] non-wood pulp production in China in 2004 (Pöyry 2006).

Traditionally, China is using a high amount of non-wood raw materials like agricultural residues (e.g. straw or bagasse) for paper production due to a lack of domestic wood feedstock. Starting in the 1950s, numerous small, rural, collectively owned paper mills were founded, with an annual production of only a few thousand tons of pulp. Missing self-owned wastewater treatment facilities, most of them caused severe water pollution. Following more strict environmental regulations, many of these have been shut down since the 1990s and have been replaced by bigger wood-based paper mills (Lang 2007). This also influences rural labour markets where around 8 million generate parts of their income from harvesting, trading or transporting biomass (Lang 2007). This trend also led to a drop of reed price and to search for new reed applications. Due to the low bulk-density, long distance transportation for reed is economically not viable and local utilisation should be favoured.

8.5 Reed use in China: Case studies

In this section, we introduce reed utilisation for four case studies: Liaohe Delta, Tarim Basin in Xinjiang, Yellow River Delta and Wuliangsuhai Lake. The Liaohe Delta and the Yellow River Delta represent two large reed areas with a continuous water supply in coastal regions. The Tarim Basin and the Wuliangsuhai Lake represent the reed use situation under arid climate in northern and northwestern China. An overview about the four case studies is illustrated in Table 4.

Table 4 – Overview about the four introduced case studies; RP = Reed price after harvest [€ /t^{-1}] (year) (Köbbing et al. 2014a).

Case study	Province	Reed area [ha]	Yields per ha [t]	Total yield [t]	RP	Reference
Liaohe Delta	Liaoning	100,000	4.5	450,000	40 (2004)	Pöyry (2006); Xiao & Li (2004); Ye et al. (2013)
Tarim Basin	Xinjiang	37,000	8	300,000	14 (2007)	Hansmann (2008)
Yellow River Delta	Shandong	100,000	15	150,000	-	Man & Croon (2007)
Wuliangsuhai Lake	Inner Mongolia	18,800	5.3	100,000	41 (2011)	Köbbing et al. (2014a)

8.5.1 Liaohe Delta

As mentioned above, the 400,000 ha Liaohe Delta (also Panjing) in northeast China (121 °10′– 122 °30′ E, 40 °30′–41 °30′ N) in the province of Liaoning is known as the biggest contiguous reed area worldwide, with an area of 100,000 ha of reed beds (Xiao & Li 2004). A sophisticated water management scheme enables people to regulate the water table according to the biological requirements of the reed beds. Reed beds are burned regularly, in order to eliminate eventual pests and other vegetation (Ji et al. 2009; Ye et al. 2013). All this had led to an annual increase of 600 ha of reed area between 1984 and 2006 (Ji et al. 2009) and levelled out the yields increases. In 2011, around 450,000 t of reed were harvested (Ye et al. 2013). The reed is harvested with traditional agriculture trucks during winter, bundled by hand, and transported by train or truck to the paper mills in the region (Ye et al. 2013).

The paper mills Yingkou Paper and Jin Cheng Paper in Yingkou and Jincheng, respectively, have an annual production capacity of 150,000 t pulp each, which corresponds to 700,000 t of reed raw material intake (calculated by a factor 2.3 from reed to pulp) (Pöyry 2006). In 2011, the two paper mills in Jincheng lacked 140,000 t reed biomass (Ye et al. 2013). In 2004, the harvested reed raw material price was € 40 /t^{-1} (Pöyry 2006), which is comparably high, probably due to the high competition between reed users. The whole management is focussed to gain high yields, other ecosystem functions and services such as habitat for birds are neglected.

8.5.2 Tarim Basin

The second case of reed use, introduced here is located in Xinjiang in the far northwest of China. Along the Tarim and its tributaries and at the Bosten Lake, reed beds are distributed. The latter has an area of 98,000 ha (Xia et al. 2001). Reed grows inside the lake, along the shores of the lake and the Tarim River and on irrigated or drainage fields. Reed at Boston Lake is exclusively used for paper production by Bohu Reed Company which is managing 30,000 ha inside the Boston Lake, plus 7,000 ha artificial planted reed at the shore (Hansmann 2008b). The reed beds are managed in terms of water table by the help of pumps and dams, in order to increase the productivity and enable an easy harvest. The harvest is done by tractors and harvesting machines or manually with sickles. The harvested reed serves as a source of cellulose for which is supplemented by 10 % wood cellulose. The paper mills has a current production capacity of 120,000 t of pulp which corresponds to 300,000 t of reed intake (Hansmann 2008b).

Apart from this one major user, reed is used in wetlands by local farmers as additional income. The reed business involves a specialised network of people dealing with cultivation, transportation, trading, manufacturing and purchasing. Final products are weaved (bora) mats (Photograph 11 and 12), reed (Yuban) panels (Photograph 13 and 14), reed blinds and bundled strings. Bora mats are sold for € 1.40 to € 2.30 and € 2.80 to € 5.60 (in 2008) depending on the quality for small (1.8 x 3 m) and big (3 x 6 m) mats, respectively. The mats are used for traditional Uighur adobe house construction. Yuban panels in contrast are produced by compressing and binding reed in one meter wide and required length panels. They were sold for € 0.40 /m² in 2008. Reed bundled in 10-15 cm rope is used for roof construction and for fencing. These bundles were purchased for € 0.04 /m in 2008 (Hansmann 2008b). A rare use of reed is in the construction of blinds for construction or as insulation material. Separate reed culms of the same length bind together in chains. The size of 2.5 x 10 m length is sold for € 2.80 in 2008 (Hansmann 2008b). Moreover, especially on sites with low productivity, reed is used as a fodder plant for sheep, goats, cattle or donkeys in spring and early summer (Gahlert 2006; Hansmann 2008b).

8.5.3 Yellow River Delta

The third reed case study introduced here is the Hekou district, Shandong Province, in the Yellow River Delta. The delta expands over an area of 780,000 ha, with reed covering more than 100,000 ha (Man & Croon 2007). With an average yield of 15 t ha^{-1}, a feedstock of 150,000 t is available (Man & Croon 2007). It is used for paper production in traditional paper mills, though, these paper mills are shut down gradually due to water pollution. New alternatives for utilisations of the reed biomass are explored, e.g. environmentally-friendly paper mills, energy production in biomass power plants or panel production.

8.5.4 Wuliangsuhai Lake

As described in Chapter 2, the Wuliangsuhai Lake is a shallow, eutrophic wetland with a dense cover of *P. australis*. An area of 18,800 ha is covered with reed, corresponding to more than half of the lake area and producing around 100,000 t of reed per year (Köbbing et al. 2014a). Harvest ensures the regular removal of biomass, which otherwise would decompose inside the lake and accelerate the eutrophication process. In this respect, a market demand for reed is important for the lake maintenance and restoration. From our investigations, Figure 2 illustrates the total lake area, reed area and yield from 1986 to 2010. In the areas bordering Wuliangsuhai Lake, roughly another 100,000 tons of reed are cultivated (Köbbing et al. 2014a).

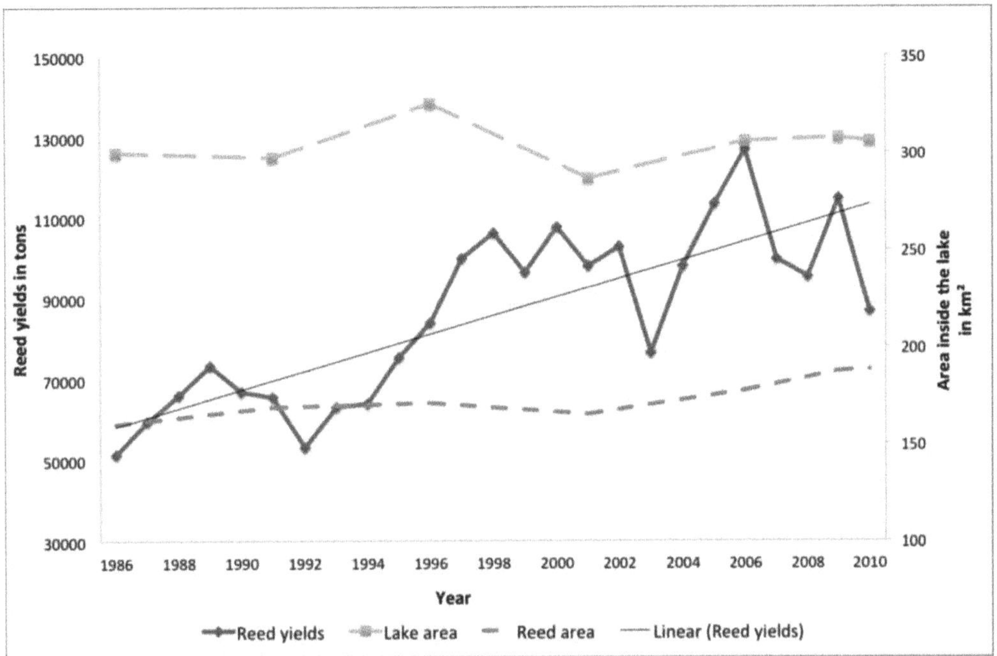

Figure 2 – Total lake area, reed area and yield from 1986–2010.

This huge biomass feedstock is an important income source for local people and for hired migrant workers. Until 2008, the reed resource was almost exclusively used for paper production in two paper mills in and near Urat Front Banner. Some minor parts were used to produce mats for construction and insulation of greenhouses (Photograph 15 and 16). More restrict environmental regulations and a missing wastewater treatment led to the shutdown of the paper mills and to two new consumers in Ningxia Province, 800 km away. The additional transportation costs had reduced the revenues of the responsible lake administration and lead to a search for new, local products.

The cut reed raw material is sold to the paper mills for € 41 /t, which includes € 22 /t paid to local fishers and farmers for the harvest. Before loading on trucks, the reed is compressed into bales (Photograph 17 and 18) which costs around € 13 /t. Transportation costs another € 20 /t which results in a reed price at gate of around € 75 /t (Photograph 19); loaded reed trucks are weighed before leaving the area (Photograph 20 and 21). The costs for pressing and transportation corresponds to the reduced income of the lake administration and also affects the income of around 4,000 people engaged in the harvesting process (Table 5).

Table 5 – Reed network and value chain at Wuliangsuhai Lake in 2011 (Köbbing et al. 2014a).

Actor	Activity	Price per Units [€ /t^{-1}]	Total [€]
Lake administration	Revenue	-	41
	Harvesting	22	-
Paper mill	Raw material	41	-
	Pressing	13	75
	Transportation	20	-

New, higher valued applications are investigated by the Wuliangsuhai Lake administration to increase profitability of the reed business. As noted earlier in this chapter, only a few reed products are suitable for large-scale applications such as panels, thatching, paper and energy. In China, reed as construction material is seen as backwards, or old-fashion, and lacks market demand. A new, clean paper mill could be an option, but requires high investments and therefore seems not to be an option.

Facing the rapid increase in energy demand in China and the negative impacts of the intensive coal consumption, reed as an energy source, as investigated in Chapter 7, offers two novel scenarios. First, the replacement of rural, coal-based furnaces by biomass ones (Scenario 1) and second, the construction of a combined heat and power (CHP) generation in a dedicated gasification plant, that is, centralised heat and power generation (Scenario 2). A 30 MW CHP plant has been considered, which corresponds to a total reed raw material consumption of 212,049 t. From Table 6, it can be seen that reed biomass furnaces can be competitive under the assumption that the energy efficiency of the new furnaces raises from 10 to 75 %. For a CHP plant, the reed price per MJ/kg will be around 1/3 more expensive. Also, the Net Present Value (NPV) valued calculated for 20 years was slightly negative. The calculation for both scenarios is based on a few assumptions. Change factors such as rising labour costs or increasing subsidies will, respectively, influence the result in a negative or positive way.

Table 6 – Heating value and prices for reed and coal at Wuliangsuhai wetland. * Inflation adjusted from February 2004 to February 2011 according the consumer price index in China reported in (OECD 2013). (Köbbing et al. 2014b).

	Present situation		Scenario 1	Scenario 2
	Rural coal	Coal plant	Reed biomass furnaces	CHP plant fed by reed
Heating value [MJ/kg]	23	23	15	15
Average purchasing price per kg * [€]	0.082	0.042	0.054	0.041
€ per MJ/kg	0.003	0.001	0.003	0.002

8.6 Conclusion

Reed as a source for multiple products has been important for China since human time. Long used as a construction material and a fodder plant, in the 1950s it was discovered as a valuable source for cellulose, which was short, due to lack of domestic forest resources. After markets opened in China the import of wood resources for paper production increased, but at the same time most of the domestic non-wood paper mills came under pressure due to their smaller size. Supply and demand, forced these smaller industries into financial despair as costs of maintaining their own water treatment facility became less supportable.

Today, reed bed managed all over China continues to innovate and change with market demands – seeking new consumer ends for reed resources. The most promising use is energy production by combustion or ethanol production, novel environmentally-friendly paper mills, natural water treatment plants and, in some cases, reed panels. But all of these products are often hardly or not profitable and need governmental support. If not, the situation will become similar to the one in Europe. Some reed beds will be reclaimed for more profitable uses like agriculture or urbanisation, others have to be costly managed (incl. harvest production) by environmental management schemes.

Key References

Dela Cruz, A. 1987. The Production of pulp from marsh grass. Economic Botany, 32: 46–50.

Köbbing, J.F., Beckmann, V., Thevs, N., Peng, H. & Zerbe, S. 2014a. Investigation of a reed economy (*Phragmites australis*) under threat: pulp and paper market, values and netchain at Wuliangsuhai Lake, Inner Mongolia, China. Technical Report. Greifswald, Germany: Institute for Botany and Landscape Ecology, University Greifswald.

Köbbing, J. F., Patuzzi, F., Baratieri, M., Beckmann, V., Thevs, N., & Zerbe, S. 2014b. Economic evaluation of common reed potential for energy production: A case study in Wuliangsuhai Lake (Inner Mongolia, China). Biomass and Bioenergy, 70, 315–329. doi:10.1016/j.biombioe.2014.08.002

Köbbing, J.F., Thevs, N. & Zerbe. 2013. The utilisation of Reed (*Phragmites australis*) – A review. Mires and Peat, 13: 1–14.

Pöyry. 2006. Technical Report, Modules 2-7. In: China: Technical Assistance for the Sustainable Development of the Non-wood Pulp and Paper Industry. Shanghai: Pöyry Forest Industry Co. Ltd.

Rodewald-Rudescu, L. 1974. Das Schilfrohr (The reed stalk). E. Schweizerbart′sche Verlagsbuchhandlung (Nägele u. Obermiller) Stuttgart.

Savcor Indufor Oy. 2006. China. Technical Assistance for the Sustainable Development of the Non-Wood Pulp and Paper Industry. Impact assessment. UNEP.

Thevs, N., Zerbe, S., Gahlert, E., Mijit, M. & Succow, M. 2007. Productivity of reed *(Phragmites australis* Trin. ex Steud.) in continental-arid NW China in relation to soil, groundwater, and land-use. Journal of Applied Botany and Food Quality, 81(1): 62–68.

Ye, S., Brix, H. & Sun, D. 2013. Large-scale management of common reed, *Phragmites australis,* for paper production: a case study from the Liaohe River Delta, China. In: International Conference "Reed as a Renewable Resource" 14-16 February 2013. 2013, Greifswald, Germany: Institute for Botany and Landscape Ecology, University of Greifswald.

9. Index of sustainable functionality: Application in Urat Front Banner

Giuseppe Tommaso Cirella
Stefan Zerbe

9.1 Introduction

The functionality of sustainable societies is a pressing notion; sustainability has, indeed, become a quintessential example of what is wrong, but at the same time embodies an ultimate practicality since it is literally meaningless unless it can be repaired. As such, it is firmly rooted in the present (Bell & Morse 2008) and in characterising its measurability one could begin investigating what is required to survive on the planet. Sustainability is an example of a paradigm recognisable from what some see as the contradictory word to sustainable growth. Paradigms are vital in that they are philosophical and theoretical frameworks within which "theories, laws and generalisations [are derived]" (Bell & Morse 2008). According to Bell and Morse (2008) the broadest spectrum of the sustainable component within the sustainability paradigm implies, and dates back to the Brundtland Report, that whatever is done now will not detriment future generations (UN 1987). However, the clear-cut definition of sustainability, and what it encompasses varies depending upon "who is using it and in what context" (Bell & Morse 2008).

In short, this chapter does not specifically focus on wetland sustainability or restoration science – it is interlinked via geography and aspects of the science and statistical data presented in previous chapters. It provides a brief look at the design of an index of sustainable functionality (ISF) model (Imberger et al. 2007; Mills et al. 2005; Cirella & Tao 2008) based on a multi-criteria analysis study of Urat Front Banner, Inner Mongolia. The research's standpoint delineates key sustainability ideas for a novel ISF formulation (Cirella & Zerbe 2014a) and, simply put, extends methodological inter-linking ideologies. Some key points from (Cirella & Zerbe 2014b) that have been identified for ISF integration include:

- uncertainty and sustainability governance, which relates to the precautionary principle and the uncertainties that come about from pathways of why we are where we are and what our carrying capacity as a species is on the planet;
- definitional concerns, noting Daly's (2006) utility- versus throughput-based notions and illustrations on a broadened spectral-view of sustainability;
- characterising measurability and natural capital, which reports on the developments of intergenerational equity concerns and conflicts between differing perspectives via constraints and thresholds; and
- measuring sustainability towards an indicator-based system, which reveals the requisite to optimise data via the use of an indicator-based system via multi-dimensional categorisation.

9.2 ISF background

The ISF of Urat Front Banner is implemented using a matrix-based model. The aim is to calculate over a 20-year period, from 1990 to 2010 in five-year blocks, Urat Front Banner's societal sustainable functionality and promote sustainable principles amongst its citizens and local authority, so it can better comprehend required action towards a sustainable way of living. The definitional status of sustainability is examined in quantitative and qualitative terms and calculated using a multi-criteria assessment. This is based on operations research in which the application uses analytical methods to help make better decisions for optimal, or near-optimal, solutions to problems. Methodology expands across a geographic domain where the related sustainability is the level of functionality within the measured area. Theoretical analysis is tested and comprised of primary research concepts, in which preliminary steps are predefined by integrating notions from previous ISF studies in betterment of formulation and mathematical controls. Triple bottom line (TBL) parameters and capital, defined as all goods that can be used in the production of other goods and services, play a key role in the definition of the methodological approach. All components of society, that being individual, community, from small business to conglomerate, and, even nation-state must ensure their actions are conducive to maintaining the life support systems that surround them (Mills et al. 2005). Hence, there is a niche for multi-faceted tools that can measure and monitor how well varying orders of life matchup against contemporary scientific standards. Many varieties of quantitative and qualitative methods of sustainability continue to be developed in an effort to transform the concept of sustainability into practical application. Key index-based examples within the scope of sustainability can be found in Table 1.

Table 1 – Key examples of sustainability-based indices based on data type.

Quantitative data type methodologies	Qualitative data type methodologies
Index for sustainable economic welfare (Daly & Cobb 1989)	Assessing the sustainability of societal initiatives and proposing agendas for change (Devuyst 1999)
Ecological footprint (Rees 1992)	Gross national happiness (Royal Government of Bhutan 1999)
Genuine progress indicator (Redefining Progress 1995)	Significance and sustainability model (Gibson et al. 2001)
Millennium development goals (UN 2000)	Quality-of-life index (Economist Intelligence Unit 2005)
Environmental performance index (preceded from the Environmental sustainability index) (Esty et al. 2005)	Happy planet index (New Economics Foundation 2006)
Living planet index (WWF 2005)	Global peace index (Institute for Economics and Peace 2007)
Index of sustainable functionality (Imberger et al. 2007)	Sustainable project appraisal routine (Arup 2008)
n-bottom line sustainability concept and performance approach (Foliente et al. 2007)	Structured analytical process for assessing measured sustainability (IUCN 2008)
Human development report (preceded from the 1990 Human development index) (UNDP 2010)	

The ISF captures components of subject complexities and acknowledges balances within its systems. Focusing upon societal assessment, typical ISF theory taps into the notion of business accounting in which a TBL approach intertwines the relationship between environmental, social and economic paradigms (Elkington 1998). The ISF differs from other methodologies as it has been formulated on a different definitional basis of sustainability. Unlike the Brundtland Report (1987), it examines the assessment of need as problematic, not as a fact but as an understanding or interpretation. The ISF overcomes this subjectivity in defining need by focusing on the present level of functionality of a system, under consideration, to indicate longevity instead of sustainability as a state or as a future point of reality (Kristiana 2009). It is in this concept that this research aspires and encapsulates continuity with other ISF-based research via a novel design and pilot study in northern China. It utilised available electronically-formulated data, and interlinked varying interdisciplinary management approaches and procedural techniques.

9.3 ISF procedural steps

The ISF geographic work is site specific and has been adapted from a number of key scientific sources and technical reports (Imberger et al. 2007; Cirella & Tao 2008; Brown & Imberger 2006; Mills et al. 2005; Kristiana 2009). The ISF framework illustrates a bottom-up approach and encompasses seven steps (Figure 1), it is theoretical and founded within the scope of a decision-aiding technique for sustainability assessment.

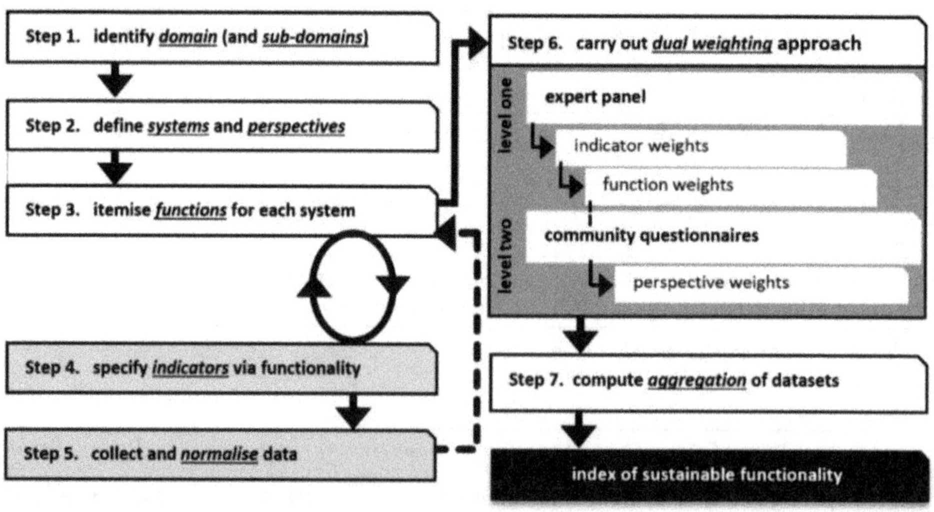

Figure 1 – ISF framework adapted from Imberger et al. (2007) and Cirella and Tao (2008).

In terms of validation, the assessment of data is mostly analytical with some parts qualitative in nature. Levels of notation, specifically functions and indicators, were selected to best suit the application via the available period permitted within the area, potential for dataset collection, expertise within the area and previous research experience and practice. The ISF formulation of Urat Front Banner has been modified from previous studies and will be briefly explained in a sequential step-by-step order. It can be reviewed in detail by reviewing (Cirella & Zerbe 2014a). The ISF equation, Eq. (1), is defined for the model. For reference purposes, associated variables are labelled throughout the procedural steps and illustrated via a list of variables in Table 2.

$$\mathrm{ISF} = \sum_{j=1}^{J^i} \sum_{k=1}^{K^i} \left[(W_{\mathrm{в}} W_{\mathrm{þ}})^{\psi} \, W^{\phi} \left\{ \sum_{l=1}^{L^i_{jk}} \frac{1}{M(\lambda^i_{jkl})} \sum_{m=1}^{M(\lambda^i_{jkl})} \left(I^i_{jklm} \right) \right\} \right] \qquad \text{Eq. (1)}$$

Table 2 – Defined variables used for the ISF formulation of Urat Front Banner (Imberger et al. 2007; Cirella & Tao 2008)

Definition		Variable
i	=	sub-domain
j	=	perspective
k	=	system
l	=	function
L	=	number of functions in the matrix element (j, k)
I	=	indicator
M	=	normalised indicator
λ	=	defined indicator range
W	=	weight
ψ	=	expert panel
ϕ	=	community questionnaire
n	=	total number
r	=	rank level
Þ	=	averaged indicator
в	=	averaged function
σ	=	averaged perspective

9.3.1 Step 1. Identify domain (and sub-domains)

The domain (D) can be broadly identified as the subject whose sustainability is being assessed. It sets the resolution of the ISF and we define N sub-domains (D^i, where $i = 1, N$) as a focused or proxy-like aspect of the domain. For this study, the domain is the geographic area of Urat Front Banner, Inner Mongolia and its sub-domains are the nine counties within the banner (Figure 2). Located in the southwest corner of the League of Bayannur, Urat Front Banner is one of seven administrative sub-divisions. It is situated on the northern bank of the Yellow River where the Hetao Irrigation District drains into Wuliangsuhai Lake. It has a total area of 7,476 km^2 and in 2010 approximately 341,600 inhabitants (Bayannur Government 2013). Population statistics show a huge influx of people between 1990 and 1995 and an almost static population since then. Table 3 illustrates this fact, labelled with numerated variables of each sub-domain (D^i), or county.

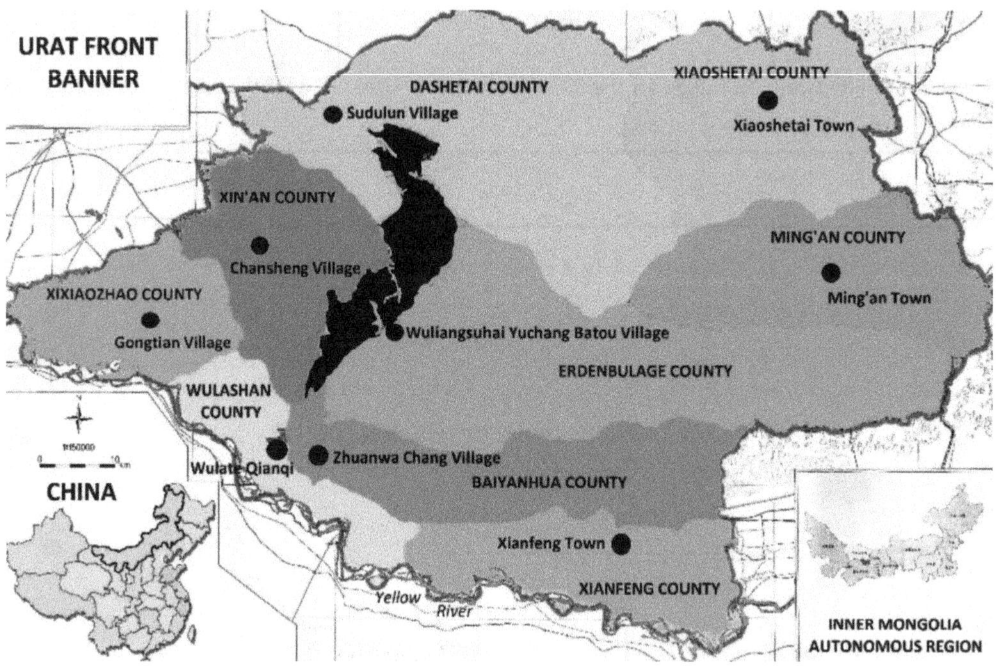

Figure 2 – The ISF domain is based on the political map of Urat Front Banner, Inner Mongolia and sub-domains correspond to its nine counties. Circles relate to a town or village, within each county, where a community questionnaire and qualitative findings were conducted (Cirella & Zerbe 2014a). Map is digitised from Wulate Qianqi Government (2012).

Historically, Urat Front Banner, since the Qin Dynasty, has been a key military station. The Hetao plain lands have acted, and continue to act, as one of the natural northern frontier regions for China. The banner has historical linkages and a strong ethnical heritage with Mongolian culture. The topography is divided between (1) fertile land around the Hetao Irrigation District and Wuliangsuhai Lake and (2) mountainous areas in the north. Key agricultural production includes wheat, sunflower (oil), melon, apple pear, tomatoes and livestock (Wulate Qianqi Government (2012), see Chapter 2 for further details on the geography of the area).

Table 3 – Population statistics of Urat Front Banner and its nine counties with labelled domain (D) and sub-domain (D^i) numerated variables (Wulate Qianqi Government 2012).

Domain (D) andSub-domains (D^i)	1990	1995	2000	2005	2010
(D) Urat Front Banner	115800	338900	330800	334000	341600
(D^1) Wulashan County	30600	105300	107500	126900	154700
(D^2) Erdenbulage County	5900	12400	11600	10800	10900
(D^3) Dashetai County	10600	46100	45300	43200	39900
(D^4) Xin'an County	12100	49500	48900	46700	41500
(D^5) Baiyanhua County	7600	13400	12300	12100	11900
(D^6) Xixiaozhao County	13600	31100	29300	26900	24300
(D^7) Xianfeng County	18600	43500	41000	36500	32700
(D^8) Ming'an County	11300	26400	24600	21400	17500
(D^9) Xiaoshetai County	5500	11200	10300	9500	8200

9.3.2 Step 2. Define systems and perspectives

For each sub-domain (D^i), we may define K^i systems S_k^i, $k = 1, K^i$. There are five systems defined, that is (1) the ecological system which is the natural environment including its components, functions and interactions; (2) the community system which is the formal and informal interactions between people, institutions and governance structures; (3) the individual system which means all human individuals who have an impact through physical and non-physical needs and attributes; (4) the economic system that is the production, distribution and consumption of traded goods and services by individuals and industry; and (5) the built system as the components which are non-living and constructed.

Perspectives are cross-referenced with systems which are intra- or inter-domain based viewpoints (Brown & Imberger 2006). They are defined as J^i perspectives P_j^i, $j = 1, J^i$ for each sub-domain (D^i). Imberger et al. (2007) annotate that this index convention configures the matrix (j, k) with J^i persperctives and K^i systems for each of the employed sub-domains. Perspectives are often influenced by the domain and are the basis for selecting functions when determining measurement.

Table 4 – Cross-sectional ISF matrix framed between systems (S_k^i) and perspectives (P_j^i), with each (j, k) box corresponding to the 15 questions from the community questionnaire (Cirella & Zerbe 2014a).

		system (S_k^i)				
		ecological	**community**	**individual**	**economic**	**built**
perspective (P_j^i)	**environmental**	**Q1.** How important is it to maintain a range of plants and animals in the countryside or natural reserves within the banner?	**Q4.** How important is it for you that the local government practices responsible environmental management?	**Q7.** How important is it for families or households to reduce their environmental impact?	**Q10.** How important is it to you that companies in the banner minimise their environmental impact?	**Q13.** How important is it to save space for a garden or agriculture by having a smaller house either on your block or within your municipality?
	social	**Q2.** How important are the parks, countryside and river in this banner for your livelihood, recreation and socialising?	**Q5.** How important is the promotion of equality and well-being within this banner's community?	**Q8.** How important is it that banner residents take an active role in the community?	**Q11.** How important is it to you that businesses in the banner pay employees enough to meet their needs?	**Q14.** How important is it to have accessible venues for entertainment in this banner?
	economic	**Q3.** How important do you think green areas are for increasing property prices in this banner?	**Q6.** How important is it for you that the local government supports local businesses?	**Q9.** How important is it to you that individuals contribute their skills to benefit the economy?	**Q12.** How important is the strength of this district/banner's economy to you?	**Q15.** How important to you is the growth of property prices in this banner?

In this study, the three perspectives are centred on the TBL approach that enable comparison on the basis of substance rather than semantics. Table 4 illustrates the matrix for this ISF analysis, which collectively is made up of the considered systems (S_k^i) and perspectives (P_j^i) – with inter-relating questions for each (j, k) point included from the community questionnaire. The matrix is an engineering viewpoint in that it functions to measure the interaction between elements of functionality (Imberger et al. 2007); it does this by evaluating each system by all the perspectives which allows for integrative sustainability assessment.

9.3.3 Step 3. Itemise functions for each system

The itemisation of functions is the point where the level of sustainability is defined; it is analogous to the notion of strong and weak sustainability. Strong sustainability is when the system can maintain all its functions and weak sustainability is when the system maintains only certain functions of choice (Imberger et al. 2007; Ott et al. 2011 see Chapter 1). This study is restricted to weak sustainability.

Defined functions F_{jkl}^{i}, $= 1, L^{jk}$, where L^{jk} are the number of functions in the matrix element (j, k). Functions define the roles and relationships performed by a specific system carried out from a particular perspective; for this study, 26 functions were selected. A sustainable state is the process when functions within a system are stable and functioning well below a critical stress threshold. It is expected that a system will rapidly degrade once this critical stress threshold is reached, as it can no longer perform its functions (Mills et al. 2005). Functions are itemised for each system by considering the three perspectives defined in Step 2. A systematic matrix of functions is created to ensure that a thorough evaluation of Urat Front Banner is performed, neglecting a minimum level of attributes. Efficiency is improved and complexity reduced as each function is itemised and cross-referenced between a relating system and perspective (Table 5).

9.3.4 Step 4. Specify indicators via functionality

The direct measurement of complex functions the ISF considers is not possible. The ISF measures the performance of functions and systems by identifying indicators via functionality. An indicator provides an indirect measurement of a system; it is only an indication of the health of that system. After a thorough literature review and consultation with local and national experts and stakeholders, 62 indicators were chosen to represent the functions itemised in Step 3. Utilising Imberger et al.'s (2007) criteria, the selected indicators must:

(1) represent the relating aspect of the function; (2) be scientifically valid; (3) be available over time, and be able to improve and decline over this period; and (4) be comparable to acceptable threshold and target levels.

Table 5 – List and categorised systems (*S*), perspectives (*P*), functions (*F*) and indicators (*I*) for the ISF of Urat Front Banner (Cirella & Zerbe 2014a).

S	*P*	*F*	*I*		(I^{i^p}) sub-indicators
Ecological system					
ecological	environmental	1		**to maintain important ecosystem processes**	
			1	water quality distributed to the local population	pH, DO_2, P
			2	water quality of the wetlands and water bodies	pH, DO_2, N, P, Cu, Pb
			3	groundwater quality	
		2		**to maintain related linkages between biodiversity of plant and animal species in an environment**	
			4	% completion of the Inner Mongolia Autonomous Region regional conservation strategy for the Hetao Irrigation Area	
		3		**to uphold climate change initiatives**	
			5	renewable energy consumption and initiatives within the region	
	social	4		**to provide aesthetic and recreational use of nature for the community**	
			6	area of green space as a % of total land area of the urban spaces	
		5		**to improve the quality of life and an environment beneficial to human health**	
			7	safe air quality levels	
	economic	6		**the provision of natural aesthetics for economic benefit**	
			8	% change in price of a property with a waterfront view, compared to % change in price of a similar property without a view	
			9	% change in price of a property next to a green space compared to % change in price of a similar property further from green space	
		7		**to provide renewable and non-renewable resources**	
			10	renewable resource – annual water consumption trends	
			11	non-renewable resource – total fossil fuels value against gross county product within the Urat Front Banner	
Community system					
community	env'tal	8		**to manage the natural environment responsibly through government policy, legislation and services – through public awareness and involvement in environmentally-friendly initiatives**	
			12	% of solid waste recycled	
			13	% of local government expenditure allocated to green spaces (Urat Front Banner)	
			14	% of local government expenditure allocated to sustainable development (Urat Front Banner)	
			15	% of wastewater reused	
	social	9		**to provide all individuals in society with equitable opportunities and outcomes – via the provision of basic services**	
			16	weekly income by gender	
			17	number of deaths per year, due to heart disease	
			18	% of annual government expenditure allocated to health (Urat Front Banner)	
			19	the number of violent and property crimes per 1000 residents	
			20	% annual government expenditure allocated to law, order and public safety (Urat Front Banner)	
		10		**to promote and encourage a diverse, interconnected and participative society**	
			21	equity of political representation	
			22	proportion of indigenous Mongolian residents, residents born overseas and Chinese born residents in the League of Bayannur, compared with the rest of Inner Mongolia	
			23	% of residents with a high school education	
			24	% of annual government expenditure allocated to education and welfare (Urat Front Banner)	
			25	% government expenditure allocated to recreation and culture (Urat Front Banner)	
	economic	11		**to support business and industry through appropriate, innovative and effective implementation of policy and mgmt strategies by the local govt**	
			26	% of annual local government expenditure allocated to economic development	
			27	average business satisfaction rating	
		12		**to provide a stable social society – via interaction and networks of interaction within the community that are oriented towards, or facilitative to, economic growth and efficiency**	
			28	the number of crimes reported per capita	
			29	the number of people who own or who have purchased (via mortgage) a home in which they are residing	

Individual system

System	Subsystem	No.	Function / Indicator
individual	environmental	13	**to maintain an environmentally sustainable lifestyle**
		30	green power consumption as a % of the total residential power consumption
		31	number of people using sustainable transport as a % of total transport use
		32	recycled residential waste as a % of total residential waste
		33	ecological footprint
		14	**to minimise waste output**
		34	total County (industry and household) waste to landfill
		35	proportion of households that participate in recycling
	social	15	**to participate in activities that contribute to the functionality of the social, political and legal system**
		36	voter turnout at local government elections (Urat Front Banner)
		37	public attendance at local government meetings (Urat Front Banner)
		16	**to take personal responsibility for own well-being**
		38	fraction of hospital related separations diagnosed with related tobacco smoking health problems
		39	fraction of hospital related separations diagnosed with related diet-related health problems
		40	emotional well-being
	economic	17	**to develop and provide human resources for production**
		41	the proportion of employed residents relative to the population in the labour force
		42	the proportion of residents with higher education qualifications relative to the employed population
		18	**to consume, save and invest to maintain economic growth and productivity**
		43	proportion of disposable income available for consumption

Economic system

System	Subsystem	No.	Function / Indicator
economic	environmental	19	**to minimise adverse impacts of industry on the environment**
		44	total energy use by industry per gross county product within the Urat Front Banner, via use of coal energy
		45	fraction of total energy use by industry that is from renewable sources
		46	industrial solid waste to landfill per gross county product within the Urat Front Banner
		47	fraction of total solid waste by industry that is recycled
	social	20	**to ensure affordability and provision of basic needs**
		48	fraction of working population whose income is over the minimum wage of the State
		21	**to create diversity and opportunity**
		49	gender equality as a deviation from 50% across all industries
		50	occupied job types as fraction of total number of possible job types in the economic System
	economic	22	**to provide accurate measures for all forms of capital which are traded in the Economic System**
		51	to contribute to economic stability and progression towards growth in the Urat Front Banner and Inner Mongolia
		23	**to contribute to stability and progression towards growth in the Urat Front Banner**
		52	county growth rate via the gross county product within the Urat Front Banner
		53	the stochastic diversity of the Urat Front Banner industry mix compared to the Inner Mongolia industry diversity
		54	growth rate of people employed in the Urat Front Banner industries

Built system

System	Subsystem	No.	Function / Indicator
built	env'tal	24	**to allow for ecosystem health**
		55	proportion of lots with deep sewerage available
		56	efficiency rating of buildings
	social	25	**to provide serviced land for societal activities**
		57	Urat Front Banner expenditure on street lights
		58	traffic accidents per resident
		59	expenditure on information technology (IT) per person
	economic	26	**to provide a network of tangible assets**
		60	increase in median house prices (i.e. the median house price is the midway point of all the houses sold at market price over a set period (monthly,yearly, quarterly, etc.))
		61	median home rental prices
		62	duration of power outages

9.3.5 Step 5. Collect and normalise data

The functionality, or degree of functioning, of each function is calculated by attributing each function a set of indicators (I^i_{jklm}) and normalising this data ($M(\lambda^i_{jkl})$) from zero to one (Imberger et al. 2007; Cirella & Tao 2008). Any value below zero will equal zero and any value above one will equal one. The normalisation of indicators is setup between upper and lower functionality limits, which transfer to bounds between zero and one (Kristiana 2009). In operational terms, one is the level at which the system is completely functional, whereas the functional bound of zero indicates dysfunctional (Imberger et al. 2007).

Linear interpolation as an approximation is employed between these two bounds. In five-year intervals dating between 1990 and 2010, data for each indicator for each sub-domain was collected (or period within that range for which data was available). Data collection is sourced from governmental reports and statistics, interviews, technical papers and relating academic literature. Data normalisation is processed in order to aggregate indicators that have different scales and units of measurement.

9.3.6 Step 6. Carry out dual weighting approach

The implementation of weight is where data is put through varying statistical weights before ISF formulation. For the ISF of Urat Front Banner, weighting of data has been used to better reflect community ownership and increase the likeliness of assessment and value (Po et al. 2003). It should be noted that this step is non-compulsory and is based upon the importance of population within the domain.

The weight of the data is determined using the dual weighting approach ($((W_{\beta} W_{\text{þ}})^{\psi}\, W^{\phi})$), first applied via the expert panel and then the community questionnaire. This correlates with Cirella and Tao's (2008) stepladder approach (Figure 3) and demonstrates a logical course of development since experts, in detail, look over the entire stepladder approach and evaluate each function and indicator, while the community, at large, only take the questionnaire that is short and general.

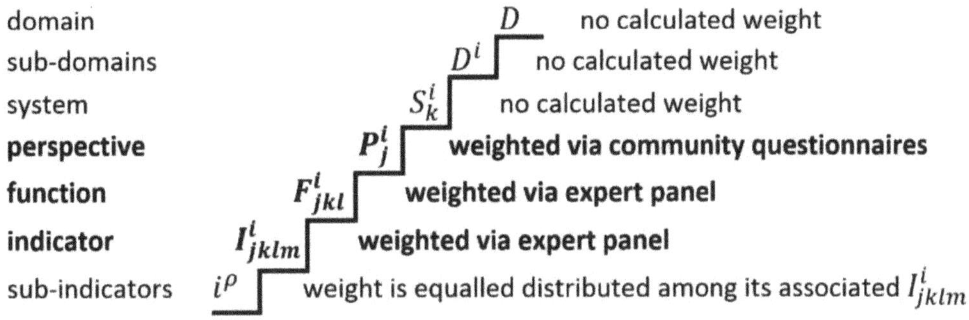

Figure 3 – The ISF stepladder approach from the lowest (i^ρ, I^i_{jklm}) to highest (D) level of notation. Calculated weight, in bold, is at the indicator, function and perspective levels (Cirella & Tao 2008).

For the expert panel, a two rounded Delphi Method of appraisal was implemented (Gordon 1994). Twenty-one experts were, first, weighted calculated indicators ($W_{þ}^{\psi}$), which calculates of indicator-to-function weight and then, functions (W_{β}^{ψ}). For the community questionnaire, one town or village in each of the nine counties was investigated between 29th September and 7th October 2013 (see Figure 2). The questionnaire included 15 questions, in which one question relates to one of ISF matrix (j, k) cross-sections, as shown in Table 4. For each county, 20 questionnaires were completed, with the exception of Wulashan County that had a sample size of 70 and Erdenbulage County that had 30. Within Wulashan County, a larger population size accounted for the extra questionnaires, while Erdenbulage County was inadvertently given more time and resources as the pilot starting point. The total number of questionnaires conducted within Urat Front Banner totalled 240, with a 95 % confidence level accordingly to target population. Specifically, the community questionnaire weighted the ISF perspectives (W^{ϕ}). This incorporates a community viewpoint by inferring more weight to higher ranked perspectives relative to that system.

The community questionnaire is weighted against the perspective-to-system relationship and completes the dual weighting approach. In addition, supplementary qualitative notes were transcribed, in each of the counties, and appraised by the expert panel members for further harmonisation via the Delphi Method. The notes focused on incorporating a better understanding of intergenerational equity concerns, conflicts between differing perspectives and harmonisation of concepts of capital and resource productivity as described in Step 2. From a viewpoint of weak sustainability, characterising measurability with experts has proven difficult and an alternative, future-based approach is also considered in Chapter 10.

9.3.7 Step 7. Compute aggregation of datasets

Aggregation uses the calculated percentile weights, or weighted sums, to compute the nine sub-domain county records via the ISF equation, Eq. (1). The ISF of Urat Front Banner is an aggregate of the county records, weighted equally as each county is considered of equal importance. This is a regional view of looking at the banner and, at the county-level, does not take into account geographical size, demographics or land-use. This self-deterministic approach, regionally, is an importance aspect of local sustainability-based thinking.

9.4 ISF analysis

The ISF of Urat Front Banner, calculated according to Eq. (1), can separated its system level (Figure 4) and perspective level (Figure 5), respectively. As these figures illustrate, a varying level of analysis can be achieved via the ISF model. The advantage of this ability is the partitioning of smaller elements and their contribution to the overall, aggregated ISF record.

For the application of Urat Front Banner, the built system shows values that are higher than other systems throughout the selected period; it also shows an increasing trend. The two social-based systems, community and individual, also show slight, consistent increases, the economic system remains mostly linear and the ecological decreases. From an overall systems viewpoint, the performance factors indicate a weak level of sustainable functionality. Important managerial concerns will need to resolve continued signs of rapid degrading natural habitat, resources, ecological balance and, to some degree, socio-economic breakdown. This may correlate with the demographic trend in population in which, basically, has remained the same since 1995. Moreover, in the last decade consistent reservations with a lack of developmental-boom, parallel with the rest of China, may be cause for concern within the banner.

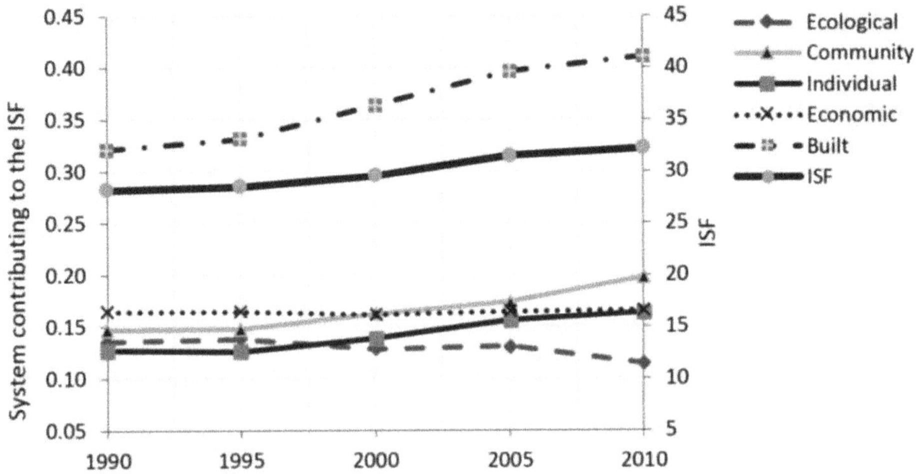

Figure 4 – ISF and system (S_k^i) functionality.

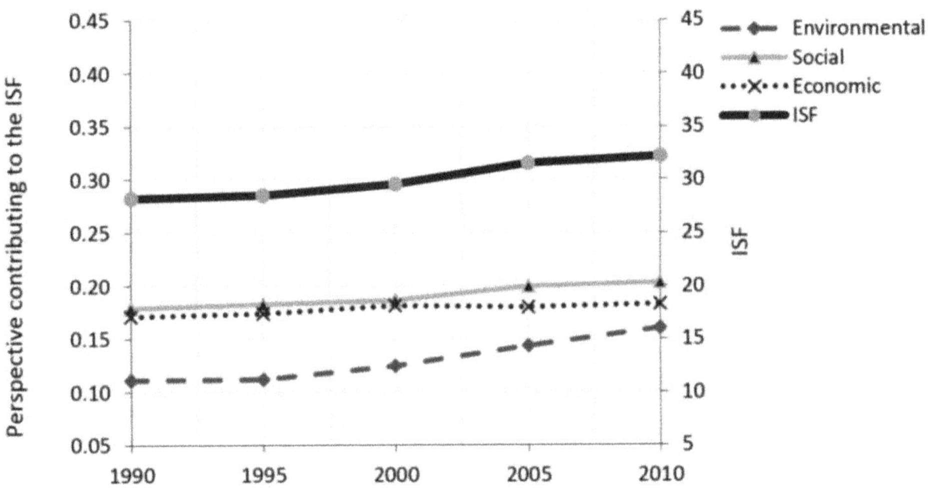

Figure 5 – ISF and perspective (P_j^i) functionality.

From a perspectives viewpoint, the appraisal shows an increasing trend in both environmental and social functionality – with economic remaining, again, constant throughout the selected period. The attentiveness, or yearning, to improve environmental functionality is positive, even though results indicate a reality that is at an insufficient level. Social perspectives are also positive with an increasing trend that may be reflective of the community and individual systems' results. Economic trends are constant and reflective of the somewhat problematic and impoverished living standards throughout the banner.

A comparative look at the banner's ISF, ecological footprint, gross banner product (GBP) and population has also been calculated (Figure 6). It shows the ISF of Urat Front Banner, composed from its nine counties, has an overall low-intermediate level of sustainable functionality.

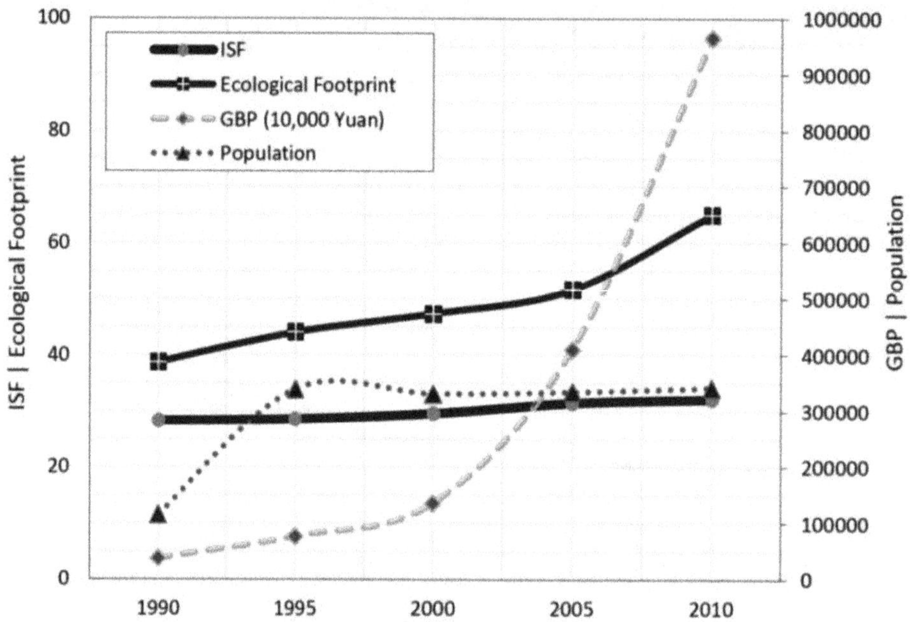

Figure 6 – Urat Front Banner: ISF, ecological footprint (estimated ranked percentage is when 100 % is equal to three planet Earths), GBP per capita and population

The record indicates three noticeable zones: (1) the two, isolated mountainous counties of Xiaoshetai County (Photograph 22) and Ming'an County which have slightly a higher ISF bi-decadal average; (2) the three counties that directly encompass Wuliangsuhai Lake of Erdenbulage County (Photograph 23), Dashetai County and Xin'an County which have a slightly lower ISF bi-decadal average; and (3) the four lower counties that are based upon the principal transportation lines in and out of the banner of Xianfeng County, Baiyanhua County, Wulashan County and Xixiaozhao County

(Photograph 24) which have analogous records to the overall banner itself. The ISF of Urat Front Banner shows a stabilised trend that is only slightly increasing over the selected period. The built system is an important aspect of the overall result and depicts the massive attempt from Chinese authorities to input infrastructure and project development into the region. This higher intermediate scoring system has bloated the overall trend from the other four low rating systems.

Comparatively, the ISF shows a much different result from the conventional economic measure of GBP per capita. Over the selected period, the GBP per capita has increased almost exponentially, especially in the last five years of the study, while the ISF only shows a minimum increase in value from 28.2 to 32.2. This indicates the increased flux of monetary growth did not translate into improved sustainable functionality within the domain. An ecological footprint was also conducted (Cirella & Zerbe 2014a) and, comparatively, shows an increase from a little over a one-planet footprint in 1990 to almost a two-planet footprint in 2010. The ISF and the ecological footprint, though calculated in quite different manners, show a general trend of low level of sustenance contrary the GBP per capita.

9.5 Conclusion

The vast amounts of data in compiling index-based research is time consuming and resource intensive. This is common to many quantitative methodologies and the nature of this study is no different. A number of qualitative attempts to investigate the possibility of adding to the dual weighting approach and formulation were investigated. The authors, continue to, suggest that the play of words, or the notion, of a 'quizzical society' is puzzle-like, and unlocking aspects of sustainable functionality exemplar, when trying to improve upon ISF resolution (Davies 2013; Ott et al. 2011). The authors construe that their previous suggestive recommendations remain underdeveloped (Cirella & Zerbe 2014b). In addition, notation development was also thought of as a possibility for improving upon the ISF model, but no concrete solutions have yet been positively tested. From a practical implementation point of view, a vast amount of metadata opens the record book for management and decision-makers at all levels. A core notion to sustainability indexing incorporates enhanced governance via historical and continual monitoring of records. Better management decisions and strong sustainability directives are future-based objectives that look at societal transitions and pretences for prospective awareness, in which, ISF trends often are not setup to measure. Utilising scenario based judgements, as noted in Chapter 1 and examined in Chapter 10, a subsequent step in sustainable thinking and consciousness towards modernity and its planned development is one such potential alternative.

Key references

Brown, A. & Imberger, J. 2006. The index of sustainable functionality (ISF): A prospective tool for assessing the sustainability of the impact of the World Bank projects. Report. Perth, Western Australia: Centre for Water Research.

Cirella, G.T. & Tao, L. 2008. Measuring sustainability: an application using the index of sustainable functionality in South East Queensland, Australia. The International Journal of Interdisciplinary Social Sciences, 3(8): 231–240.

Cirella, G.T. & Zerbe, S. 2014a. Index of sustainable functionality: Procedural developments and application in Urat Front Banner, Inner Mongolia Autonomous Region. The International Journal of Environmental Sustainability, In press.

Cirella, G.T. & Zerbe, S. 2014b. Quizzical societies: A closer look at sustainability and principles of unlocking its measurability. The International Journal of Science in Society, 5(3): 29–45.

Daly, H.E. 2006. The Future of Sustainability. M. Keiner ed. Dordrecht, The Netherlands: Springer.

Foliente, G., Kearns, A., Maheepala, S., Bai, X. & Barnett, G. 2007. Beyond Triple Bottom Line – Sustainable Cities: CSIRO. In: State of Australian Cities National Conference (SOAC2007). 2007, Adelaide, Australia:

Commonwealth Scientific and Industrial Research Organisation, 28–30 November.

Imberger, J., Mamouni, E.D., Anderson, J., Ng, M., Nicol, S. & Veale, A. 2007. The index of sustainable functionality: a new adaptive, multicriteria measurement of sustainability – application to Western Australia. International Journal of Environment and Sustainable Development, 6(3): 323–355.

10. Development scenarios on the Hetao Irrigation District and Wuliangsuhai Lake

Lilin Kerschbaumer
Konrad Ott
Niels Thevs

10.1 Introduction

The ecological changes occurring within the area of Wuliangsuhai Lake is closely linked with the agricultural development in the Hetao Irrigation District (HID). From our research, we argue that it is highly unlikely to preserve the Wuliangsuhai area as a lake without realising substantial change in the HID (see Chapter 2 for background information of this site). The shortcomings of previous work within Wuliangsuhai Lake and its surrounding area is a result from an isolation of the lake area, in which it does not take into account the causal nexus between the HID and the critical state of the lake itself. To clarify those causal linkages and to envision the prospects of Wuliangsuhai Lake in the future, qualitative scenarios (i.e. narratives or story lines) have been developed by the authors of this chapter, which have based different sustainability paradigms, proposed over a time span of 30 years. The general intention of our scenario writing was to sharpen awareness of local stakeholders regarding different possible routes of development and their probable outcomes. Scenarios generally postulate that the future is open to some degree and that present decisions and actions have impacts for better or worse. The different sustainability paradigms that underlie the scenarios include:

- intermediate sustainability (*scenario A*): conservation of "critical natural capital";
- strong sustainability (*scenario B*): maintenance and gradual increase of each kind of natural capital based on the Constant Natural Capital Rule (CNCR);
- weak sustainability (*scenario C*): maintenance of total capital (i.e. sum of natural capital and human-made capital); and
- non-sustainability (*scenario D*): worst case where a wasteland appears.

Figure 1 refers to the notion of natural capital which is presupposed in all of the four scenarios. The general notion is arranged as a scheme which comprehends different types of natural capital.

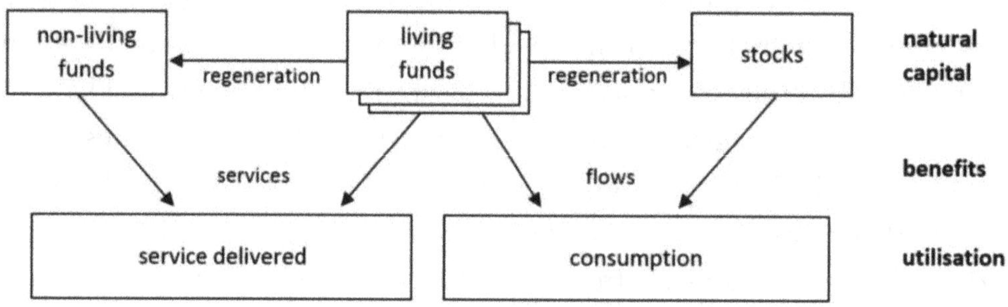

Figure 1 – Scheme of natural capital, based on Ott and Döring (2008).

"Natural capital" comprises of all those components of nature that bring some benefit or enrich the many capabilities of human beings and higher developed animals or serve as preconditions for utilisation (Ott & Döring 2008). The concept of natural capital refers to the many ways and modes on which nature has beneficial impacts on human (or animal) life. "Natural" does not mean that nature is free from any human interference (e.g. "wilderness"). It includes stocks and funds which have been modified by human action, but remain natural to some degree. Thus, we mostly face "cultivated natural capital" in Europe and China, for example.

The notion of benefit is not restricted to economic welfare but encompasses non-material benefits as well (e.g. recreation, aesthetics and the like).

"Critical natural capital" enables the provision of essential ecological functions and ecosystem services. Natural capital can be divided into stocks and funds (as described in Figure 1). Stocks yield benefit streams, while funds deliver services. Unlike stocks, funds regenerate or reproduce themselves; therefore, fund services continue over time if the respective fund is not depleted. Hence, the utilisation rate of such funds should not exceed their regeneration rate, if those service flows are to be maintained. One main cause of over-exploitation and depletion of natural capital is to perceive funds as if they were stocks. Different from natural capital, "human-made" capital cannot be found in the nature but must be produced. It includes production factors (e.g. equipment, input materials and so on), human capital, intellectual capital and social capital, etc. The sum of natural capital, human capital and human-made capital compose the total capital a given society holds at a given time or a given period of time. The different paradigms, therefore, propose different utilisation schemes on capital. In a nutshell, weak and intermediate concepts of sustainability allow for more substitution between different kinds of capital while strong sustainability is more demanding with respect to conservation and restoration of natural capitals, especially funds. A shift from the mere depletion of stocks to the restoration of funds is strongly required by strong sustainability (*scenario B*). Strong sustainability takes into account funds of cultivated natural capital. By doing so, the problem of substitutability eventually returns within this concept. Our scenarios address this problem.

Freshwater is a non-living fund under the above outlined scheme. In the HID and Wuliangsuhai Lake area, freshwater resources includes those from the Yellow River (diverted from the Sanshenggong Water Station) and those in Wuliangsuhai Lake. Figure 2 shows the consumption level of such natural capital. After the peak in the 1990s, water consumption in the HID and Wuliangsuhai area dropped below 5 billion m^3/yr (Bm^3/yr), which is similar to the level before the 1980s. The recent drop in water consumption reflects a stricter implementation of the water use quota regulation noted in the "Yellow River Water Resource Allocation Plan" (NDRC & MoWR 1998), however water consumption level still exceeds its assigned quota of 4

Bm³/yr. Freshwater is a highly critical natural capital in the region. Other types of natural capital include air, soil, forests, grasslands, coal and minerals. It is worth noting that natural capital like coal and minerals are stocks, which mean they can only be depleted and do not regenerate themselves on a human timescale. Once used up, their benefit streams stop and only their external effects remain (i.e. debris, waste and greenhouse gas emissions). At present, the HID and Wuliangsuhai Lake area depends heavily on such stocks of natural capital for energy generation and gross domestic product growth.

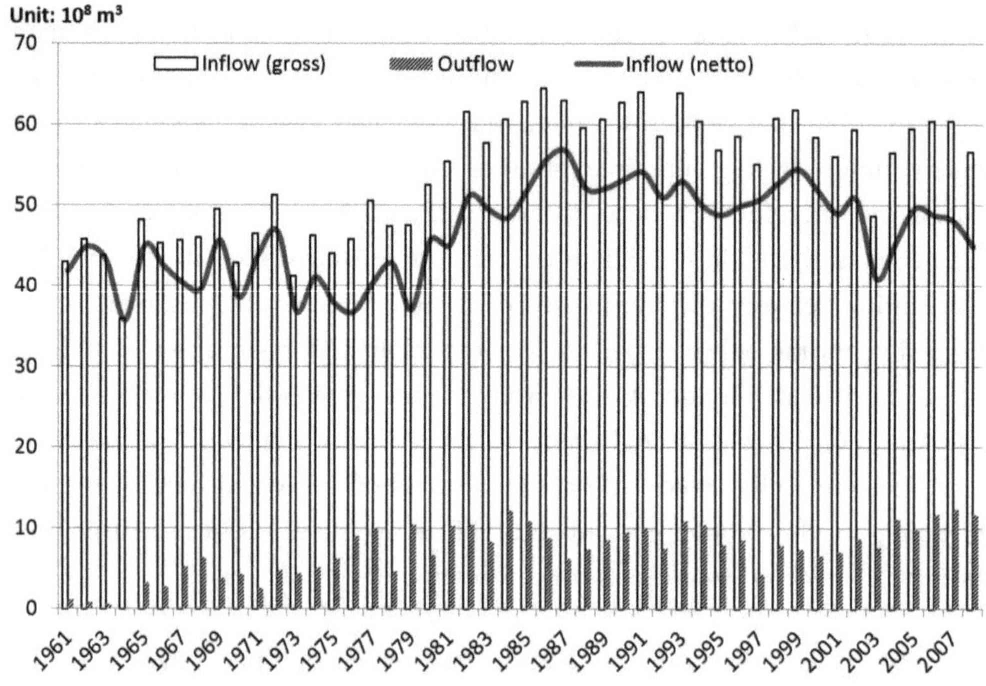

Figure 2 – Water resources in the HID and Wuliangsuhai Lake area. "Inflow (gross)" is the water amount diverted at the Sanshenggong Water Station, "Outflow" is the water amount returning to the Yellow River and "Inflow (netto)" is the linear overhead of the "Inflow (gross)" subtract the "Outflow" (BCPG 2010).

The details of the four scenarios are outlined in Table 1. In the first three scenarios, the back-casting technique has been applied. It asks in each scenario the question of *"What is to be done today to reach a specific sustainability goal?"* The fourth scenario highlights the possible consequences of a non-sustainability, "worst case" situation. Since those consequences are to be avoided, there exists no goal in this scenario and the narrative is based on forecasting instead of a back-casting technique.

Behind the interaction between agricultural development and ecological changes in the HID and Wuliangsuhai Lake area, there are crucial drivers of changes (e.g. institutional re-arrangement, agro-technological advancement and environmental transformation). Initiating those changes are governmental authorities, actors in the private sector (e.g. industry) and in the agricultural sector (e.g. farmers and fishers) and non-governmental organisations (NGOs). Their action orientations differ greatly. Governmental authorities focus on securing food output that not only meets regional basic needs but also brings stable economic income through export to other regions throughout China. This could further contribute to social cohesion in the HID and Wuliangsuhai Lake area. Private sector actors include production agents outside the primary sector, comprising of agro-product processing enterprises. Their major concern is the return of investment. The locals (i.e. actors in the agricultural sector) include mainly farmers and fishers. They aim, as we shall assume, to maximise their incomes within a given institutional and technological framework. Lastly, local NGOs (i.e. Water User Association (WUA) and the Boya Cultural Association (BCA)) act on behalf of disadvantaged and under-represented groups (e.g. human beings of current and future generations, animals, landscapes, etc.). They are in close cooperation with other stakeholders. The stakeholder situation being given at HID and Wuliangsuhai Lake make it likely that short-term interest may, as often, dominate policy-making. On the other hand, however, stakeholders are searching for viable pathways out of a critical and inconvenient situation and actually hope to avoid the loss of Wuliangsuhai ceasing being a lake.

It is worth noting that *scenario A* can transit to either *scenario B* or *scenario C* with additional measures. Of course, *scenario B* and *scenario C* can also be developed from the baseline situation (see Chapter 5). During a project meeting in Linhe, Bayannur in 2011 local authorities and experts identified the region to be in the transition from *scenario A* to *scenario B*. However, the reality shows otherwise. With expanding areas of submerged vegetation and increasing frequency of algae boom, the lake is on the verge between a grass type and an algal type ecosystem. Its fishery and tourist activities have basically come to a halt.

Table 1 – Scenarios for the HID and Wuliangsuhai Lake area, adopted from Kerschbaumer et al. (2014).

scenario A intermediate sustainability	*scenario B* strong sustainability	*scenario C* weak sustainability	*scenario D* non-sustainability
goal			
- conserving Wuliangsuhai Lake as a lacustrine ecosystem with an open water area remaining at ca. 293 km²	- developing Wuliangsuhai Lake into a Ramsar wetland of international importance under CNCR in the HID and Wuliangsuhai Lake area	- developing Wuliangsuhai into a palustrine ecosystem; expansion of a reed economy and establishment of sustainable tourism	- no goal, Wuliangsuhai Lake turning into a wasteland
measures (based on individual initiators)			
governmental authorities (focus: food security, economic prosperity and social stability)			
- diversion of extra water (0.56 Bm³) from the Yellow River - water quota at 4 Bm³/yr; cultivation area at 5740 km² - promotion of innovative technologies like non-irrigation afforestation technology and water harvesting afforestation technology - promotion of organic fertilisers and enforcement of phasing-out old fertilisers - subsidisation of organic farming to farmers - establishment of a local fund for restoring Wuliangsuhai Lake - restoration projects in Wuliangsuhai Lake	- decrease of cultivation area to less than 5740 km²; regional land use based on Man and Biosphere - enforcement of organic farming during the take-off period (5–10 years) - setting incentives for agro-product companies to purchase organic crops from farmers at a fixed price - assistance programs to teach farmers required techniques (in cooperation with NGOs) - assistance programs to market organic agro-products from the HID (in cooperation with NGOs) - restoration projects in Wuliangsuhai Lake, including water diversion from the Yellow River	- incentive programs and policies to expand a reed economy - assistance programs to market reed products from the HID (in cooperation with NGOs) - construction of wastewater treatment plants in case of excessive nutrient discharge from the HID - establishment of a tourist industry in Wuliangsuhai Lake	- higher possibility of extreme events impacting area, especially the agricultural sector - high costs to construct more wastewater treatment plants - health threat to people in the region - threat of drinking water safety, especially to downstream regions - loss of income source due to disappearance of the fishery, reed or tourist industry - loss of cultural significance of the HID and Wuliangsuhai Lake area

private sector actors (focus: return of investment)		
- business as usual - participation in the local fund to restore Wuliangsuhai Lake and in Water Rights Transfer projects	- participation in the local fund and Water Rights Transfer projects -investment in natural capital	- production of diversified reed products (see Chapter 8)
agricultural sector actors (focus: income maximisation without welfare jeopardisation)		
- new plantation scheme consisting of wheat, proso millet, sorghum, grain legumes and maize - return of suitable stalks of grain to the fields as fertiliser	- new plantation scheme consisting of wheat, proso millet, sorghum, grain legumes and maize - return of suitable stalks of grain to the fields as fertiliser -active participation in WUA -sustainable fish yield	- HID: business as usual - Wuliangsuhai Lake: fishers shifting to reed harvesting (in winter) and tourist industry (in summer)
NGOs (focus: protection of disadvantaged groups)		
- assistance in establishing a semi-water market - assistance to farmers in their transition to organic farming	- assistance programs to teach farmers required techniques (in cooperation with governmental authorities) - assistance programs to market organic agro-products from the HID (in cooperation with governmental authorities)	- assistance programs to market reed products from the HID (in cooperation with governmental authorities)

The severe environmental degradation in Wuliangsuhai Lake prompted reports and commentaries from major Chinese news agencies in 2012 on possible disappearance of the "Pearl of the Northern Frontier" in 20 years (Wu 2012; Li & Zhang 2012). The HID and Wuliangsuhai area is more likely to be on a path towards scenario C and, perhaps, scenario D. Scenario B, however, places strict requirements on the region to change their business-as-usual practices. Without sufficient political will, its realisation would be impossible. The next section argues that *scenario B* is most desirable and still feasible but implies short-term opportunity costs.

10.2 The case for *scenario B*

Scenario B adopts a holistic view, asking for the maintenance and gradual increase of natural capital in the whole region (i.e. CNCR), and not just a particular kind of natural capital in a specific location. Freshwater in Wuliangsuhai Lake is no doubt a critical natural capital in the region. However, other kinds of natural capital should not be left aside or sacrificed, like forests and grasslands. Just as in the fishery sector in Wuliangsuhai Lake, the utilisation of forests for commercial wood and grasslands for husbandry is to follow the concept of principle of sustainable yield (i.e. the harvesting rate not exceeding its reproduction rate). Responsible governmental authorities should cooperate closely with each other for sustainable use of respective natural capital, instead of chasing individual GDP performance. This means for some or all governmental officials a departure from their respective "comfort zones", and a relatively radical shift in the routine and logic of interactions with each other on the individual, as well as, departmental level.

Secondly, expanding organic farming poses great challenges to all stakeholders. Despite governmental policies promoting organic farming, organic agriculture does not yet have sufficient competitive advantages in China, as a whole, compared with traditional agriculture. This lies in, above all, the underdeveloped system of the organic agro-industry in all operation stages, including production (especially the educational level of farmers) and sales and certification versus insufficient informational transparency in the whole of the industry (Chang-wei et al. 2010). The development of organic farming in Dengkou County, Bayannur illustrates exactly this issue in which the project faced a dilemma after financial support from its cooperation partner ceased. The taking-off of organic farming (i.e. in the first 5–10 years), as *scenario B* requires, is to be enforced by governmental authorities. The enforcement should not lower the welfare level of farmers or related parties in the agro-production chain. Without *scenario A* as transition in which organic farming is promoted, the pressure on governmental authorities becomes even higher in terms of investment in infrastructure and financial assistance to farmers and agro-product enterprises during the taking-off period. Governmental assistance for capacity building should continue even beyond the take-off period, say, for at least 30 years (i.e. the proposed time span of the scenario study).

Furthermore, the constraint on the farming area, which totals 5,740 km^2 within the HID and Wuliangsuhai Lake, at the same time puts a limitation on agricultural output. Even if organic farming might be as productive as traditional farming, a decrease in total output might be highly possible due to farmers being unfamiliar with the new praxis. Sustaining the welfare level of related parties in the agro-industry appears to an extreme challenging task for the government. For farmers, the new praxis plus new plantation schemes bring uncertainty in terms of their income. This could be one of the major obstacles to enforce organic farming if farmers insist on their routine praxis and are unwilling to shift to the alternative path. Figure 3 shows the dominating cultivation of melons and tomatoes since about the early 1990s, followed by sugar and oil plants. It takes, therefore, time to shift this routine cultivation praxis to a new one. Stable policy, informational transparency and successful examples could release such concerns. It is essential for farmers to take active part in the decision-making process, for example through respective WUAs. This requires a corresponding educational level as well as a local culture to promote democratic management. At the moment, the above mentioned seems still to be lacking far behind.

Thirdly, commitment of industry to invest in natural capital is difficult to establish. Industrial participation in Water Rights Transfer projects is in most cases determined by related policies, and is not an option if new industrial projects require extra water. Similarly, it is possible to implement the local fund through governmental regulations. Also, there could be regulations to restrict the establishment of industries with high water demands or with high potential of water pollution.

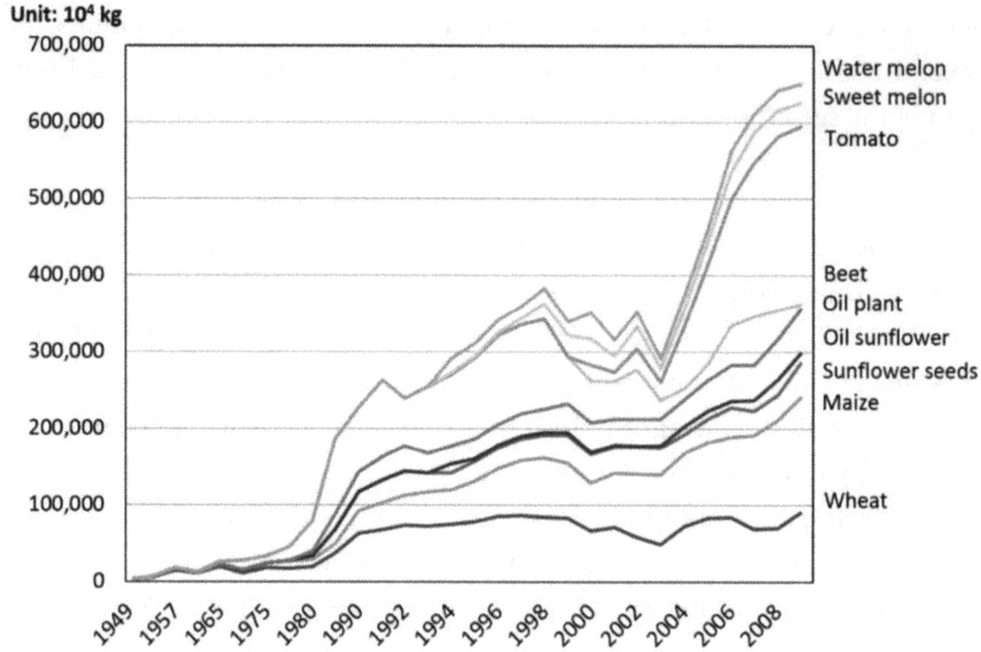

Figure 3 – Agricultural output of major crops in the HID between 1949 and 2009 (BCAH 2013).

Difficult to reach are binding commitments for the investment in natural capital from the industrial sector; that the industrial sector is a major contributor to regional GDP growth endows them with certain power on the negotiation table, which quite often rends environmental protection regulations void. If CNCR is to be held, the widespread guideline of "polluting first, cleaning follows" in existing policies and regulations needs to be discarded. However, since pollution discharge fees are a major income source for environmental protection authorities, there lacks incentives for a change. Shifting the function of environmental protection authorities is a long-term task that requires bold political reforms. Furthermore, even though activities to protect the environment might improve corporate image, enterprises in the market economy, especially small and medium sized ones, have the foremost goal of winning profit. Profit maximisation is frequently prioritised over environment and social responsibilities. In a situation where a solid corporate social responsibility (CSR) system is missing, as it is at present, achieving strong sustainability looks highly unrealistic.

Fourthly, influential NGOs are missing. In the HID and Wuliangsuhai Lake area, existing NGOs include various WUAs and the BCA. The former focuses on water issues, while the BCA focuses on the protection of Wuliangsuhai Lake mainly through training programs. As such, the limited focuses within a constrained political framework make it difficult for NGOs in China as whole to fully function. In their cooperation with governmental authorities or with industry, they quite often find themselves being directed rather than initiating direction. Nonetheless, strong sustainability asks for active roles from stakeholders in an equal manner. NGOs represent, above all, disadvantaged groups. Therefore, their constrained influence means limiting voices of those groups. However, if strong sustainability is to be achieved, the voices of NGOs should be taken seriously. It requires the independence of NGOs, and hence changes in the political culture in China. It is, of course, possible to start those changes bottom up, that is, on the local level. However, there still lacks political will in the Bayannur local government in this regard, or better said, the issue is not even on the agenda.

In conclusion, the management guideline of "polluting first, cleaning follows" runs exactly in the opposite direction of strong sustainability. It allows depletion of natural capital for the purpose of achieving growth in human-made capital. Restoring after polluting is very difficult in such a semi-arid region like the HID and Wuliangsuhai Lake area. End-of-the-pipe solutions will only help in the short-term. They cannot halt environmental degradation in the whole of the ecosystem. On a larger framework, such management guideline violate the principle of "Maintaining the Healthy Life of the Yellow River" (HLR) put forward by the Yellow River Conservancy Commission (see Kerschbaumer and Ott (2013) for a detailed discussion on the HLR approach).

Wuliangsuhai Lake is an integral part of the Yellow River. HLR, hence, applies on a regional scale. The "health" of the Yellow River requires a healthy Wuliangsuhai Lake. What, then, does "healthy" mean? The Wuliangsuhai Comprehensive Treatment Plan ("the Plan") by BCPG (2010) sets out the following two goals:

1. On the short-term, i.e. 2008–2015: to maintain water level, ensure ecological water demand, reduce inflow pollutants by 50 %, maintain water quality at Level V according to the national "Environmental Quality Standards for Surface Water (GB3838-2002)" (MoEP & QSIQ 2002), bring down the extent of eutrophication and paludification, increase biodiversity level, and reduce soil erosion;

2. On the long-term, i.e. until 2020: to bring pollutant sources and soil erosion under control, further reduce the amount of inflow pollutants by 20 %, enhance water quality to Level IV, and complete improvement of the aquatic ecosystem.

Just as proponents of the HLR approach, the initiators of the plan adopt a functionalist understanding on "(ecosystem) health". However, it should be noted that the notion itself entails requirements on both the natural properties as well as social properties of concerned ecosystems – "a dialectical unity" as HLR proponents argue. The plan, therefore, inherits partiality in its goals. It needs to take social properties, particularly cultural and spiritual values, into consideration as well. The HID and Wuliangsuhai Lake area is a highly diversified society in terms of ethnic cultures. The ethnic groups of Han (i.e. Daoist and Confucianist traditions), Mongols and Tibetans (i.e. Buddhist traditions) have their own cultural perception on water and on what should be considered ethical in utilising it. These social aspects are important if strong sustainability is to be placed. After all, hydrological problems are "wicked", where "true-false" solutions are not possible but "better-than" ones (Norton 2005). Cultural values play an important role in searching for "better-than" solutions.

From an institutional perspective, *scenario B* follows the CNCR rule, that is, combined with other widely recognised and applied principles, like those from the Convention on Biological Diversity (CBD) and the Ramsar Convention on Wetlands (Hu & Ge 2004; RCW 2005). CBD's central guiding principles are "the conservation of biological diversity, the sustainable use of its components, and the fair and equitable sharing of the benefits arising out of the utilisation of genetic resources" (UN 1992). Ramsar promotes the wise use of wetlands, defined as "the maintenance of their ecological character, achieved through the implementation of ecosystem approaches, within the context of sustainable development" (RCW 2005). The above principles need to be made operational on the local level in the HID and Wuliangsuhai Lake area. This is only possible with a strong political will to reform the current institutional structure and sufficient capacity of all parties involved to participate in practical discourses. All in all, contrary to what local authorities and experts expect, the HID and Wuliangsuhai Lake area is still far away from achieving strong sustainability, but appears rather to be nearing *scenario D* where Wuliangsuhai Lake turns into a wasteland. Besides the possible consequences outlined in Table 1, Wuliangsuhai Lake is likely then to undergo a process of salinisation, becoming potentially a source of sandstorms. *Is there a way out?*

The answer can be "yes". The proposed set of four scenarios should not be considered as a "finished" set. Scenario analysis as a technique in future studies aimed at expanding the visions on a given issue and supporting decision-making. In the case of the HID and Wuliangsuhai Lake area, the basic scenarios are developed in order to set off discussions among stakeholders. Measures proposed in each scenario are subject to negotiation and argumentation. During the dynamic decision-making process, stakeholders can combine various elements in the basic scenarios to form a new set of scenarios. For stakeholders in the HID and Wuliangsuhai Lake area, the above discussed challenges come with different degrees of difficulty with varying time-span requirements. For instance, the taking-off of organic farming practices might borrow experiences from successful cases (e.g. like in Dengkou County), and therefore would be relatively easier than institutional reforms to establish a democratic decision-making system in the region.

As said above, stakeholders are free to establish new scenarios after thorough considerations of various components in the basic scenarios. It is essential that all stakeholders participate with equity in this process, and not just the experts and authorities. The way out from a wasteland scenario towards a way in to a sustainability scenario is still possible if all those involved work together towards meeting the challenges outlined above. Institutional re-arrangement, corporate social responsibility and capacity building play critical roles in achieving this goal. The following are some concrete suggestions:

Institutional re-arrangement

- setting clear regional development goals under the strategy to "establish a harmonious society" and to "construct ecological civilisation" put forward by the central government;
- establishing a fair benefit-sharing system through legislations;
- improving policy implementation especially regarding environmental assessment and payment for ecosystem services;
- replacing GDP growth as standard to evaluate governmental performance with sustainability indicators;
- improving inter-departmental cooperation among governmental units;
- strengthening the role of NGOs in the regional decision-making process;
- increasing financial support (both from public expenditure and private investment) to capacity-building for locals, especially the disadvantage groups;
- raising awareness in all walks of life and society for environmental values; and
- establishing a culture of open dialogue.

Corporate social responsibility (CSR)

- establishing CSR in agro-product enterprises to promote responsible water use and "*wu gong hai nong chan pin*" (agro-products without public hazards); and
- establishing CSR in the industrial sector for responsible water use in terms of their impacts on regional water quantity and water quality and in terms of commitment to invest in regional natural capital.

Capacity building

- skills and techniques for organic farming praxis in HID (farmers);

- capacity for sustainable fishery and tourism within Wuliangsuhai Lake (fishers and agents in the tourist sector);
- awareness for environmental impacts; and
- capacity to actively participate in practical discourses.

10.3 Conclusion

In principle, people are free to choose any scenario. This freedom of choice, however, should not be confused with moral and political autonomy. While freedom of choice, as in consumerism, might rely merely on given interests and preferences, moral and political autonomy rely on reasons which presume to transcend narrow egotism with respect to the common good. As we have argued in the previous section, there are such reasons that clearly speak in favour of *scenario B*. The concept of ecological civilisation which has become prominent in contemporary China, is highly coherent with *scenario B*, while it is clearly incompatible with *scenario D*. As we have argued elsewhere (Kerschbaumer et al. 2014), *scenario C* implies large risks since it might be collapse into *scenario D* if boundary conditions, as climate, may change. As we have argued, *scenario B* is desirable, feasible, and viable.

Desirability can be substantiated in terms of sustainability but the term "Pearl of the Northern Frontier" implies that the loss of such "pearls" is undesirable.[1] The tragedy of contemporary China might be that it is losing its natural "pearls" and receives some commercial goods and infrastructures in return. This is "weak sustainability". Feasibility can be substantiated with respect to institutional change and new incentives in HID. Change does not come about by itself but it requires political action, prudent transition management and even leadership. In situations of crisis, persons matter. Viability can be predicted with some confidence. As we know from many cases in protected area management, initial resistance of local people often

1 This is a conceptual remark we hold to be true by conceptual implication. If a natural item is termed "pearl", then it is valued as being precious. It is obvious that it is undesirable to lose something that is precious. It is incoherent to argue, "X" is highly precious to me but I do not care at all if I lose "X". See Ott (1997, Ch. 2) where such implications are further elaborated upon.

changes into acceptance if the new regime has been established and can realise its benefits.

The still existing "window of opportunity" to reach *scenario B* is, however, gradually closing and it would be practically impossible to reach the targets of *scenario B* if *scenario C* or even *scenario D* have come into existence. Current decision-making has to face this kind of irreversibility. If so, all stakeholders and authorities cannot deny their moral and political responsibility since they are clearly facing disaster and evil.

Key references

BCPG. 2010. Wuliangsuhai comprehensive treatment plan (Wuliangsuhai Zonghe Zhili Guihua). Bayannur, China: Bayannur City People's Government.

Kerschbaumer, L., Köbbing, J.F., Ott, K., Thevs, N. & Zerbe, S. 2014. Development Scenarios on Hetao Irrigation Area (China) – A qualitative analysis from social, economic and ecological perspectives. Environmental Earth Sciences, DOI: 10.1007/s12665–014–3061–8.

Kerschbaumer, L. & Ott, K. 2013. Maintaining a River's Healthy Life? An Inquiry on Water Ethics and Water Praxis in the Upstream Region of China's Yellow River. Water Alternatives, 6(1): 107–124.

Norton, B.G. 2005. Sustainability: A Philosophy of Adaptive Ecosystem Management. Chicago: University of Chicago Press.

Ott, K. & Döring, R. 2008. Theorie und Praxis starker Nachhaltigkeit. Marburg: Metropolis-Verlag.

11. Conclusion and interdisciplinary recommendations

Giuseppe Tommaso Cirella
Lorenzo Brusetti
Marco Baratieri
Ping He
Niels Thevs
Konrad Ott
Stefan Zerbe

11.1 Internationalisation: A healthier planet

On a global scale, water problems stem from our failure to meet basic human needs, ineffective institutions and management and an incapability to balance human needs with the needs of the natural world. These difficulties are imbedded in a wasteful use of water, characterised by poor management systems, underinvestment and unseemly economic incentives, failure to apply existing technologies and an outdated mind-set focused almost wholly on developing new centralised supply chains. This is to the exclusion of conservation strategies in which the delivery of water-related services matches user needs and resource availability (Pacific Institute 2014). Our interdisciplinary research team compliments the growing worldwide trend of focusing on efficiency and restoration measures, at all levels, to create sustainable communities and a healthier planet.

This book has presented a number of examples of exploring sustainable water management and sound strategies for wetland restoration. In a combined effort, we have specifically researched these two topics, over the past four years, from a multi-scientific standpoint. These two topics have predominantly grown in-parallel and, in anticipation of our findings have indicated otherwise. We have learnt, to some degree, that from an interdisciplinary point of view they touch base and, their relationship, begins to show characteristics that intertwine at a conceptual or theoretical level. These characteristics indicate a cohesive past, present and future-based manner of reasoning, and point toward a societal transition to better understand sustainable action and its relating pathways.

Sustainable water management primarily looks at water as one of the basic needs for survival. Literally, water is life due to the fact that virtually no species can survive without it. While, wetland restoration focuses on the rehabilitative capability of such land that has been degraded or destroyed. For both topics, land preservation and conservation are a key to establish a harmonised viewpoint and a supportable generational-friendly future. Interdisciplinary recommendations for a sustainable water management scheme – inclusive of wetland restoration strategies – is developed throughout this chapter. A brief look into northern China examines the significance and sets the tone for the recommendations that follow.

11.1.1 Insight: Water crisis in northern China

As noted throughout this book, in northern China there is a growing pressure on water resources. This pressure effects environmental well-being, social parity and economic progress. Northern China is not immune from its nation's rapid infrastructure development, pollution concerns or population challenges – as engagement programs throughout much of the north, and west for that matter, continue to date (Wu & Ci 2002; Dowling & Wignaraja 2006; Zhou et al. 2008; UNDP 2012). China, as a whole, is still modernising itself like many other wealthier countries did in the past half century – by growing first and, hopefully, cleaning up later. According to a UNDP (2012) report on sustaining human progress, the world cannot afford a China that follows this model. Mechanisms that can assist in alleviating this challenge include (1) agricultural production (i.e. changes primarily in growing crops, raising livestock, land use and deforestation); (2) fair and balanced consumption; (3) design and creation of greener cities and technologies; (4) raising rural resilience; and (5) support for cleaner energy generation (i.e. to reduce or eliminate pollution). To this end, northern China faces critical challenges that go well beyond water management and wetland restoration. However, these two topics offer some first-step insight or potential resolution to the growing concerns since issues of this magnitude are connected (i.e. from local to global in scale and structure).

11.1.2 Regional transparency: Central Asia

We contend the geographic study areas, based in northern China, have a degree of transparency with much of the neighbouring regions of Central Asia. This is due to the nature, or mosaic, of the landscapes, wetland environments and industry-related conditions region-wide (Kreuzberg 2005; Perelet 2007; Thevs et al. 2012). The extent of our recommendations, therefore, though specific to northern China, are not exclusive. Bordering Central Asian countries with China are especially pertinent as they share direct trade routes and important cultural and historical linkages. Moreover, the recommendations have a backdrop of complexity and diversity ranging from somewhat rural, poverty-stricken to booming-like urban municipalities. Northern China, and parts of Central Asia (as noted in Chapter 2), face this type of development crisis (Zhou et al. 2008; Zerbe & Thevs 2011; UNDP 2012).

11.2 Interdisciplinary recommendations: The science

The science behind our recommendations are based on rigorous scientific experimentation and are verifiable via methodologies that can be reproduced and tested. Much of the work has been published in international peer-reviewed scientific journals. The basis of this book and our research has followed this format as closely as possible, with little exception. The extent of this book is not to write policy, but to be a tool for decision-makers and management personal, in the study areas, working within the field of sustainable water management and wetland restoration.

The state-of-the-art of typical shallow macrophytic lakes in northern China (including Wuliangsuhai Lake, Baiyangdian Lake and Bosten Lake) is that they are at risk of shrinking or even disappearance. The shrinking of these water bodies is mostly due to human interference by way of resource utilisation. The disappearance combines the human factor, just mentioned, with reed and submerged plant overgrowth, which is due to eutrophication of the water and plant residual deposits. The key to keeping northern China's macrophytic lake areas healthy is the continual purification of the water quality by aquatic plants or by harvesting and utilising them. The basis of developing a sustainable water management scheme is to understand the science and to continue to monitor the issues.

The recommendations, first, start with a brief look at the complex political issues which demonstrate the difficulties of maintaining healthy wetlands and reed production. Second, an examination of anti-pollution based research, by way of environmental indicators, help strategise wetland restoration and conservation practice. Common reed, used as an example and key species throughout much of this book, is an important resource and component to sustainable living patterns in which local people are the beneficiary. Third, economic uses of reed are exemplified through different utilisation of plant stocks. Fourth, in conclusion a holistic viewpoint is argued.

11.3 Wetlands in China: Political frontiers

The political processes in China are complex, making it difficult to pinpoint solutions in an environmentally-stricken country. This is mostly due to rapid development by way of a strong economic and political drive. These political actions are not sustainable; infrastructure and economic goals are prioritised. The environmental ethics of comprehending the basic question "*Why preserve nature?*", argued in Chapter 1, debates this by stating human beings and future generations are and, most likely, will be dependent on the ongoing utilisation of nature (as a resource, reservoir, sink and medium) since certain states of it bring feelings of joy, pleasure, well-being, peace and delight. This understanding of nature puts forth a paradigm in which scientists, themselves, are challenged to uphold and prototype this basic question in their work. The SuWaRest project is no different. Working within China, scientific rules or norms, are mostly construed to data-mining based research. This poses somewhat of a problem when authorities are questioned or issues of political correctness arise. Local people are under a moderate amount of pressure not to intervene with any level of authority. This political clout needs to be looked at closer, if wetland environments, or any natural habitat for that matter, are to be conserved and preserved anywhere in China. Placing this argument aside, wetlands throughout China have had some national attention and conservation planning over the last few years – stating their importance and ecological benefits as key driving factors.

According to China's State Forestry Bureau (2014), the National Wetland Inventory Survey between 2009-2013 indicated the area of wetlands within Qinghai Province, Tibet Autonomous Region, Heilongjiang Province and Inner Mongolia exceed five million ha, which accounts for about 50 % of the total wetland area in China (i.e. 53.6 million ha). As a vital ecosystem, wetlands perform key ecological functions in these arid, semi-arid and semi-humid regions in which they conserve important water resources for down-stream users. Specific to this book, wetlands in Zhangye, Gansu Province and Wuliangsuhai Lake, Inner Mongolia represent two of the typical wetlands in the arid and semi-arid regions of northern China.

A large number of swamp wetlands exist in piedmont alluvial-diluvial fans adjacent to Tianshan Mountain, Qilian Mountain and Taihang Mountain from where groundwater is withdrawn and river discharges flow (UNESCO 2006). Along these passes there is a relatively high abundance of water resources. Cities are often densely located flanking alongside these piedmont areas. Some examples include cities like Zhangye, Jiuquan and Wuwei that are located in the piedmont of Qilian Mountain. Similarly, an economic zone composed of ten or more cities can be found along the piedmont of Tianshan Mountain where cities including Urumqi, Shihezi and Karamay are located. Finally, some large and medium-sized cities like Beijing, Baoding, Shijiazhuang, Xingtai and Handan are located along the piedmont of Taihang Mountain. In China, piedmont wetlands shrink and can even disappear due to damming of mountain passes. They also frequently suffer from groundwater overuse which can leave entire cities without water for extended periods (Appleyard 2007; Chen et al. 2008; Hubacek et al. 2009).

Unfortunately, the degradation does not end there, it has been reported that residual wetlands are often polluted from urban wastewater discharge (Lu et al. 2009; Jolivet et al. 2014), as noted by Borruso (2014) in Zhangye.

As a result of this over-development, the Chinese government has increased its attention towards wetland protection. Scientific research is one of the aspects these initiatives incorporate. Its goal of expanding wetland areas, as stated at the 18th National Congress of the Communist Party of China, is part of an important national ecological program (Hu 2012; MOFCOM 2012). A redline for wetland areas measuring 53.3 million ha has been put forward in a plan for the promotion of ecological progress (State Forestry Bureau 2014). In theory, as a result, the area of wetlands stated should be preserved until the year 2020 (i.e. all present wetlands should be reclassified as park land). This corresponds with a number of Chinese scientific studies that state any heightened period of development due to both human activity and climatic change, have an urgent need to budget key ecological systems and services, such as wetlands (i.e. rice paddies and natural wetlands) and lakes (i.e. inclusive of reservoirs and ponds), which are sensitive to these changes (Yuan et al. 2014; Feng et al. 2013; Chen et al. 2013; Wang & Liang 2013). Apart from national initiatives, from a research standpoint, the shrinkage of wetland areas due to overgrowth and residual sedimentation should immediately be monitored for any change in water quantity or functional level of water quality (i.e. water degradation).

At the national level, China has put forward the idea that the development of ecological progress should be combined with political, economic, social and cultural means. At present, there is a plan to develop a close relationship between ecological problems and society, economy, culture and institutions – and research, such as potential reed economies, the well-being for peasant farmers, water resource utilisation and pollution discharge from an environmental and ethical aspect be closely considered (Hu 2012). It focuses on solutions that integrate the perspectives of society, economy and culture – and research ideas and achievements to provide an important reference point for ecological progress being carried out China-wide (State Forestry Bureau 2014).

11.4 Anti-pollution: Environmental indicators and reed stand-based research

The two study sites investigated in northern China offered a vast assortment of data and results from a multi-array of scientific perspectives. In Zhangye, we find a modern city on its way to developing as an impressive industrial and tertiary-based conurbation. Zhangye's wetlands face problems with industrial wastewater pollution. While, the Hetao Irrigation District and Wuliangsuhai Lake are agricultural areas that have potential at becoming important recreational zones for ecotourism. They face serious concerns of eutrophication and salinisation. At large, these points are prime hurdles for the regions' environmental management and advancement. Both areas, unfortunately, have an intensified level of pollution due to industrial and agricultural activities. In consequence, a key question for local administrators, when dealing with these issues, is the assessment of its water purification systems. This can be better understood with the use of reed stand-based research. Much of the work within this book is formulated on these observations and their potential use of environmental indicators and relating linkages.

Ecosystem quality of wetlands, covered within Chapter 5, found that plant nutrient stoichiometry can be utilised as an indicator for ecosystem characteristics. Plant nitrogen and phosphorus stoichiometry research in contrasting reed ecotypes (i.e. from aquatic and terrestrial environments), concludes that reed can perfectly adapt to differing conditions with distinct leaf and root functional traits. Isometric biomass allocation patterns are also key. Suggestive resource acquisition strategies suggest that reed can be used as an important bioindicator for understanding and managing reed dominated wetland ecosystems. This conclusion complements the following recommendation from our microbiology studies.

11.4.1 Microbiological research: The use of bioindicators

Our microbiological research illustrates how bacterial communities are powerful bioindicators for different types of pollution. In Chapter 3 and 6, a better understanding of what kind of pollution, historically, effects the two study areas, as noted in Chapter 2, is disclosed using freshwater sediment examination. To date, this type of pollution assessment is one of the best methods when looking at past pollution data.

First, in Zhangye, nutrients like nitrogen and phosphorous are important study-pollutants in urban areas, together with cadmium and mercury, whose concentrations are strong enough to affect bacterial community diversity and composition. In industrial zones, several metals are very widespread, increasing the probability of dangerously impacting upon ecosystems and human population. The effects of these different pollutants decrease in agricultural areas and in the Heihe River. There is a significant presence of genes conferring antibiotic resistance. This genetic trait widely correlates with antibiotic use (i.e. probably from hospitals and clinics, but also husbandry) and metal ions pollution. Consequently, a quite diffused pollution level continues to influence the city area of Zhangye, even though a newly installed wastewater treatment plant ameliorates freshwater quality for urban use.

Collected reed stand samples from a nearby national park appear to be cleaner than other samples from around the city. Hence, it can be deduced that the phytoremediation potential of *P. australis,* and of its associated bacterial communities, are structurally similar to those from the cleanest areas of Heihe River Basin. Moreover, it is noteworthy to state that a strongly eutrophic pond on the outskirts of Wuliangsuhai Lake, in the middle of a highly polluted industrial area, contains a bacterial community quite similar to its adjacent agricultural areas (Borruso 2014). Since eutrophication is known to have an excess amount of nutrients, *P. australis*-associated bacterial communities that are located within typical cleaner areas, indicate that *P. australis* and other macrophytes are capable of water purification even in a very adverse environment. This is a positive point for

Zhangye's management and environmental advisory. On the west side of the industrial zone there are extended lowlands, only partially used as pastures. These areas can be used as a natural and extended reed phytoremediation plant zone that include already existing reed stands and could be adjoined as park lands or, similarly, as recreational areas.

Second, considering the Hetao Irrigation District, adjacent with Wuliangsuhai Lake, our focus has been on the main channel and one secondary drainage channel that leads up to the lake itself. Our analyses show a general good environmental quality, especially in the western part of the main drainage channel. We found this part of the channel with a vast coverage of *P. australis* stands. The presence of reed-covered areas decreased from the middle part of the main drainage channel, mainly due to a larger basin and livestock grazing. In this area, additional water from other drainage channels, some of that from nearby industrialised towns, decreased slightly due to freshwater sediment quality, without dramatically affecting overall ecosystem levels. The secondary drainage channel has been more influenced by human activities, since villages discharge their urban wastewater directly into the channel. A weak antibiotic resistance gene signal is evident in the middle part of the main drainage channel, signalling a worrisome concentration of antibiotics already in the environment. Our recommendation is to prolong the potential use of *P. australis* stands in cleaning these water systems. The growth and harvesting of reeds could also be extended along the main drainage channel, leaving free localised areas to allow for sheep and other livestock space water access. Together with a greener industrial behaviour, this will lead to an increase in water quality and decrease of Wuliangsuhai Lake's ecological stress. On the long-term, our recommendation is to continually monitor the areas and introduce an environmental impact assessment plan. Such a plan would become a key for environmental policies and should help administrators alike choose better strategies to increase health and future sustainability practices.

11.4.2 Wetland restoration strategies in northern China

Wetland restoration is a process that assists in transforming a degraded wetland area that has been impacted by human activities in order to re-establish certain ecosystem services (e.g. self-purification of water, reed production, tourism and recreation). The process might be long-term oriented and requires a detailed understanding of the environment. This includes the historical consideration of the area and its succession through time to better help decide how such a restoration process and its management are to be conducted. Success can be ascertained if the wetland area can recapture its natural dynamics and original ecosystem services. However, it is impossible to refashion ecosystems back to a virgin state, as nature itself is made up of the changing entropic elements that structure our planet.

Different types of wetlands require different strategies. In northern China, we have examined stream corridor and open aquatic lake systems in which fresh water environments exist. Differing restoration strategies depend on the degradation level (Timmermann et al. 2009). After having carried out restoration measures, a continual monitoring is crucial in detecting concerns before they develop into something unmanageable. Monitoring will involve biological, geological, hydrological, chemical and physical components of the wetland areas. Specific to northern China, wetland strategies and management should take into account material risk; that is, the restoration process should not have a material adverse impact on the accessibility of water, safeguarding of biodiversity, employment, local peoples and community or land access for agriculture.

Ecological concerns and physical restrictions are imperative when a scheme for wetland restoration or creation is planned (Lüderitz et al. 2010). Wetland restoration can nowadays be based on many decades of experience (VanAndel & Aronson 2006; Lüderitz et al. 2010). According to Kentula (2002), destroying the function of an existing wetland, or other ecosystem, in exchange for another wetland function encompasses numerous questions such as *"Which is more important, the existing or the replacement function?"*,

"Will the proposed wetland increase wildlife diversity?" and *"Is the increased diversity worth the loss of habitat of any endangered species?"* Questions of this manner always should be asked during the planning stages of any wetland restoration and creation blueprint. Zeng et al. (2012), Kusler and Kentula (1990) and Kentula (2002) have all reported wetland restoration as more of an art than a science; that is, its functional proxy has not been overwhelmingly corroborated. In northern China, the science and engineering of wetland restoration has two key probable factors that limit its effectiveness for a successful wetland restoration project: (1) lack of data on ecologically mature restored and created wetlands, and on the maturation process; and (2) the limited number of well-devised and well-constructed project wetlands that can be used as exemplar prototypes (Kentula 2002). Generally speaking, restoration is prospectively going to be more successful than a full-scale creation. Within the two study sites, the restoration of damaged or destroyed wetland areas will have a greater chance of establishing a range of prior wetland functions, if a monitoring program is incorporated. Our recommendation is a long-term persistent scheme in which the restored wetlands facilitate the potential use of *P. australis* reed stands in cleaning the waterways and wetland ecosystems. With this in mind, we briefly will touch upon the varying reed production in northern China and recommend it as a future-based resource for wetland restoration and ecosystem quality of such wetland restoration settlements.

11.4.3 Reed production: A resource for the future

With our long-term recommendation for restoring the studied wetlands with common reed, *P. australis*, a community-based viewpoint should be integrated with an economic one. For a restored wetland to co-exist primary reed production must have an outlet in which local communities and businesses can strive to survive on. The primary reed production areas of China are located in northeast China, northwest China (Xinjiang) and Inner Mongolia. Broken down, the large areas of reed are found in the Liaohe River Delta, Songnen Plain and Sanjiang Plain of northeast China; Bosten Lake, Ili Valley and Emin River Valley in Tacheng of Xinjiang; Baiyangdian Lake and Hengshui Lake of the North China Plain; and Hulunbeier

Grassland and Xilin Gol Grassland of Inner Mongolia. At the moment, the level of industrialised reed exploitation in these areas remains quite low (Köbbing et al. 2014a). Most of the reed is used for paper and straw mats (as stated in Chapter 8). Additionally, biomass technology has not yet been kick-started due to a lack of economic and inefficient utilisation of reed (Patuzzi et al. 2013a). Currently, the biomass approach of burning straw generates a large amount of air pollution and causes great concern for the environment.

Desperately needed research in reed resource utilisation – beyond the traditional use of it in China as fodder, mats, baskets, huts, construction material and fire starting material (Hansmann 2008a) – has more recently moved towards large-scaled paper production plants (see Chapter 8). However, even though reed is managed all over China, its most promising uses are its usability as an energy source (i.e. via combustion or ethanol), within environmentally-friendly paper mill manufacturing, natural water treatment plants and, in some cases, reed panels. Unfortunately, at the moment, none of these usages are economical and encroaching agricultural lands and urban and suburban limits continue to hamper sustainable water management and wetland restoration planning. An immediate recommendation, in accordance with the Report of the 18th National Congress of the Communist Party of China's (Hu 2012; MOFCOM 2012) wetlands initiative, is to implement this initiative as soon as possible, by way of state-wide preservation and conservation methods. In prospect of this accord, the following years should be ample time to implement a viable usage for reed production, while at the same time creating and implementing a stronghold wetland restoration program.

11.5 Economic costs and benefits for different reed utilisation

As stated previously in this chapter, reed is a plant with multiple functions and services in which many, to date, have been overlooked. Commercial utilisation of reed can raise awareness regarding the importance of wetlands, but should be also analysed regarding profitability under market conditions. In remote areas of northern China, to be competitive against cheap coal at the household or power plant level, reed biomass feedstock has to be exploited locally – to limit transportation costs – and subsidised by the government. According to our economic analysis performed, the evaluation of energetic use of reed under the scenario of local decentralised heating and large-scale cogeneration plants, even a 30 MW power plant could barely be profitable under current conditions. Presently, China's energy policy has focused primarily on large-scale power plant industry, many of them in the developed southeast. Small, decentralised projects, which are important for remote rural areas lack support. One reason for this shortfall from renewable energy is the pre-existing energy implementation law, in which refusal to the energy grid by big energy companies or technical handlings demonstrate the difficulties and irregularities for reed-based renewable energy. The dominance of state, coal-based energy prevents the growing of alternative decentralised – mostly renewable energy – bottom-up approaches.

Nevertheless, conditions may change. On the one hand, decreasing investment costs due to a high rate of innovation and increasing energy prices may positively influence the net positive value of combined heat and power production plants. Whereas, on the other hand, labour and transportation costs may augment due to the economic growth and rising energy prices. This would negatively affect the net positive value. Considering variable reed costs, a balanced benefit-cost relation for a 30 MW plant can be achieved if the reed price decreases, within the region, by 5 CNY/t (approximately € 0.60/t) to 365 CNY/t (approximately € 42.90/t). From the perspective of reducing greenhouse gas emissions, local utilisation should be favoured over long transportation distances. Benefits can be possibly increased by accounting the greenhouse gas emission mitigated by use of biomass energy under the Kyoto Protocol. Clean Development Mechanisms (CDM) in the United Nations Framework Convention on Climate Change are implemented to allow developing countries reducing their greenhouse gas emissions by financing mitigation projects in such countries. Each reduced ton of CO_2-equivalent can be traded as "Certified Emission Reduction" (CER) on the international carbon market. Renewable energy projects play only a minor role in CDM which are mostly restricted to larger scales, due to high transaction costs. Unfortunately, the price for CER trading experienced a rapid decline in recent years, accounting only for € 0.24/CER in November 2013.

Using the example of Wuliangsuhai Lake, if prices increase again, CDM could play a role in financing a reed cogeneration plant. Considering that the CO_2-emission factor for coal is 93 kg/MJ, an amount of 86.7 kg/MJ and 260,000 t of CO_2 could be saved considering the present and potential harvesting of reed in a cogeneration plant of 10 MW and 30 MW. In addition, considering a distributed energy cogeneration scenario suitable to generate the same two overall power levels, 92.5 kg/MJ and 309,700 t of CO_2 would be saved, respectively. Furthermore, non-market values could also be considered in a holistic management scheme in and around the Wuliangsuhai Lake area. Regular harvest of reed removes a considerable amount of biomass from the lake, which prevents a rapid silting and "second pollution" by way of decomposition. Also considering local employment during the winter season, reed cutting plays an important role, currently employing up to 2000 workers. Hence, the economic costs and benefits for different reed utilisation further supports our recommendations, for long-term restoration, of planting and utilising common reed in northern China's wetlands.

11.6 A holistic standpoint

The potential debates on possible water management schemes for the two research sites opens the argument to what has happened in the past, what currently is in practice and what future scenarios can be thought up. Management schemes should be sustainable and should include ideologies that embed local cultures. Our examination of the different "water cultures" has looked at sustainability concerns via differing methodologies (Chapter 9 and 10). Within a holistic viewpoint, water in settlements and agriculture have been intertwined within the SuWaRest project's interdisciplinary fields. They include (1) site, vegetation and restoration; (2) phytoremediation and water quality, including microbiological aspects; (3) energy production; (4) cost-benefit analysis of reed use; and (5) water culture and sustainability. From a holistic viewpoint, our recommendations are an alteration in the concept of water perception, value, attitude and utility at a societal scale. Successful conservation practice requires legal regulation but it also touches upon the background of culture itself, its habits, customs, traditions, modes of perception and framings. This is especially accurate within the merging highly diverse cultures that are site-specific, found throughout much of Central Asia. The use of technical skills in conjunction with managerial-based socio-economic thinking is an urgent matter. Environmental improvements, using up to date environmental standards, must continue to be validated and should ethically be substantiated, ideally, by way of local communities actively participating in decision-making processes. This would be a starting point towards a fully holistic outlook on sustainable water management and a better threshold point for the relating-settlements' wetland restoration.

Sustainable water management indicates a cohesive need to better fully understand environmental, social and economic aspects at play. This especially includes the alternative scenarios, as discussed in Chapter 10, which exemplifies future-based modelling from where it can take us to how it may fully function. Our final recommendation, and visionary goal, is for a strong sustainability of the two study sites. The existing opportunity to achieve such a goal, however, gradually is closing and will become virtually unachievable if continued weak sustainability, even worse, non-sustainability targets are reached. Our international and interdisciplinary team has strived to derive our recommendations in a hope that waters and wetlands in northern China, and beyond, adopt a sustainable management plan that has a holistic viewpoint for these, ever so, vital ecosystems.

Contributors

Marco Baratieri, Dr.
Faculty of Science and Technology, Free University of Bozen·Bolzano
Piazza Università 1, I-39100 Bozen·Bolzano, Italy
marco.baratieri@unibz.it

Luigimaria Borruso, Dr.
Faculty of Science and Technology, Free University of Bozen·Bolzano
Piazza Università 1, I-39100 Bozen·Bolzano, Italy
luigimaria.borruso@natec.unibz.it

Lorenzo Brusetti, Dr.
Faculty of Science and Technology, Free University of Bozen·Bolzano
Piazza Università 1, I-39100 Bozen·Bolzano, Italy
lorenzo.brusetti@unibz.it

Giuseppe Tommaso Cirella, Dr.
Faculty of Science and Technology, Free University of Bozen·Bolzano
Piazza Università 1, I-39100 Bozen·Bolzano, Italy
gtcirella@gmail.com

Ping He, Prof. Dr.
Institute of Ecology, Chinese Research Academy of Environmental Sciences
Dayangfang 8, Bei Yuan Road, Chaoyang District, Beijing 100012, China
heping@craes.org.cn

Henrike Hochmuth
Institute for Botany and Landscape Ecology, University Greifswald
Soldmannstr. 15, D-17487 Greifswald, Germany
henny_hochmuth@freenet.de

Lilin Kerschbaumer
Department of Philosophy, Christian-Albrecht-University of Kiel
Leibnizstr. 6 | Zi. 316 24118 Kiel, Germany
lilin.yu@gmx.net

Jan Felix Köbbing
Institute for Botany and Landscape Ecology, University Greifswald
Soldmannstr. 15, D-17487 Greifswald, Germany
jan.koebbing@stud.uni-greifswald.de

Liping Li, Dr.
Faculty of Science and Technology, Free University of Bozen·Bolzano
Piazza Università 1, I-39100 Bozen·Bolzano, Italy
liliping415@163.com

Konrad Ott, Prof. Dr.
Department of Philosophy, Christian-Albrecht-University of Kiel
Leibnizstr. 6 | Zi. 316 24118 Kiel, Germany
ott@philsem.uni-kiel.de

Francesco Patuzzi, Dr.
Faculty of Science and Technology, Free University of Bozen·Bolzano
Piazza Università 1, I-39100 Bozen·Bolzano, Italy
francesco.patuzzi@natec.unibz.it

Niels Thevs, Dr.
Institute for Botany and Landscape Ecology, University Greifswald
Soldmannstr. 15, D-17487 Greifswald, Germany
niels.thevs@uni-greifswald.de

Stefan Zerbe, Prof. Dr.
Faculty of Science and Technology, Free University of Bozen·Bolzano
Piazza Università 1, I-39100 Bozen·Bolzano, Italy
stefan.zerbe@unibz.it

References

Abramowicz, D.A. 1995. Aerobic and anaerobic PCB biodegradation in the environment. Environmental Health Perspectives, 103: 97–99.

Ács, É., Borsodi, A.K., Makk, J., Molnár, P., Mózes, A., Rusznyák, A., Reskóné, M.N. & Kiss, K.T. 2003. Algological and bacteriological investigations on reed periphyton in Lake Velencei, Hungary. Hydrobiologia, 506–509(1–3): 549–557.

Allen, H.K., Donato, J., Wang, H.H., Cloud-Hansen, K.A., Davies, J. & Handelsman, J. 2010. Call of the wild: Antibiotic resistance genes in natural environments. Nature reviews. Microbiology, 8(4): 251–259.

Allirand, J.-M. & Gosse, G. 1995. An above-ground biomass production model for a common reed (Phragmites communis Trin.) stand. Biomass and Bioenergy, 9(6): 441–448.

Altimira, F., Yáñez, C., Bravo, G., González, M., Rojas, L.A. & Seeger, M. 2012. Characterization of copper-resistant bacteria and bacterial communities from copper-polluted agricultural soils of central Chile. BMC Microbiology, 12(1): 193.

APHA. 1995. Standard Methods for the Examination of Water and Wastewater. 19th Ed. Washington, DC: American Public Health Association.

Appleyard, B. 2007. Smart Cities: Solutions for China's Rapid Urbanization. Natural Resources Defense Council.

Arias, B., Pevida, C., Fermoso, J., Plaza, M.G., Rubiera, F. & Pis, J.J. 2008. Influence of torrefaction on the grindability and reactivity of woody biomass. Fuel Processing Technology, 89(2): 169–175.

Arup. 2008. SPeAR: Product overview. [Online]. 2008. Available from: http://www.arup.com/ [Accessed: 6 October 2013].

Atchison, J.E. 1995. Twenty-five years of global progress in non-wood plant fiber pulping—historical highlights, present status, and future prospects. In: Proceedings of the 1995 Pulping Conference, 1–5 October, Chicago, IL, vol. 1. Atlanta, GA: TAPPI Press, 1995: 91–101.

Bai, F., Li, W.P. & Li, Z.H. 2008. Analysis on the main causes resulting in vegetation degeneration in the Heihe River Basin. Arid Zone Research, 25: 219–224.

Bais, H.P., Park, S.W., Weir, T.L., Callaway, R.M. & Vivanco, J.M. 2004. How plants communicate using the underground information superhighway. Trends in Plant Science, 9(1): 26–32.

Balaji, D.S., Basavaraja, S., Bedre Mahesh, D., Prabhakar, B.K. & Venkataraman, A. 2009. Extracellular biosynthesis of functionalized silver nanoparticles by strains of Cladosporium cladosporioides. Colloids Surf B, 68: 88–92.

Baldantoni, D., Alfani, A., Di Tommasi, P., Bartoli, G. & De Santo, A.V. 2004. Assessment of macro and microelement accumulation capability of two aquatic plants. Environmental Pollution, 130(2): 149–56.

Baran, M., Váradyová, Z., Kráâmar, S. & Hedbávn, J. 2002. The Common Reed (Phragmites australis) as a source of roughage in ruminant nutrition. Acta Vet. Brno, 72: 445–449.

Baratieri, M., Baggio, P., Fiori, L. & Grigiante, M. 2008. Biomass as an energy source: Thermodynamic constraints on the performance of the conversion process. Bioresource Technology, 99(15): 7063–7073.

Barnett, T.P., Adam, J.C. & Lettenmeier, D.P. 2005. Potential impacts of a warming climate on water availability in snow-dominated regions. Nature, 438: 303–309.

Barton, D.N. 2005. Economic analysis of the value of water in alternative uses in the Lake Wuliangsuhai catchment. Inner Mongolia lake restoration project, NIVA.

Baruch, Z. 2011. Acta Oecologica Leaf trait variation of a dominant neotropical savanna tree across rainfall and fertility gradients. Acta Oecologica, 37(5): 455–461.

Barz, M., Ahlhaus, M. & Wichtmann, W. 2006. Energetic Utilization of Common Reed for Combined Heat and Power Generation. In: 2nd International Baltic Bioenergy Conference. 2006, Stralsund, 168–175.

Bayannur Government. 2013. Wulate Qianqi Year Report: 2013. Bayannur, China.

BCAH. 2013. Statistics on crops production Bayannur City Animal and Husbandry Bureau. [Online]. 2013. Bayannur City Agriculture and Husbandry Bureau. Available from: http://www.bmagri.gov.cn/Pages/DataCenter/CropList.aspx [Accessed: 12 April 2012].

BCPG. 2010. Wuliangsuhai comprehensive treatment plan (Wuliangsuhai Zonghe Zhili Guihua). Bayannur, China: Bayannur City People's Government.

Bedard, D.L., Ritalahti, K.M. & Löffler, F.E. 2007. The Dehalococcoides population in sediment-free mixed cultures metabolically dechlorinates the commercial polychlorinated biphenyl mixture aroclor 1260. Applied and Environmental Microbiology, 73(8): 2513–2521.

Bell, S. & Morse, S. 2008. Sustainability indicators: Measuring the immeasurable? 2nd Ed. London: Earthscan.

Benson, D.R. & Silvester, W.B. 1993. Biology of *Frankia* strains, actinomycete symbionts of actinorhizal plants. Microbiological Reviews, 57(2): 293–319.

Berendsen, R.L., Pieterse, C.M.J. & Bakker, P.A.H.M. 2012. The rhizosphere microbiome and plant health. Trends in Plant Science, 17(8): 478–486.

Berg, J., Tom-Petersen, A. & Nybroe, O. 2005. Copper amendment of agricultural soil selects for bacterial antibiotic resistance in the field. Letters in Applied Microbiology, 40(2): 146–151.

Bergman, P.C.A. & Kiel, J.H.A. 2005. Torrefaction for biomass upgrading. In: 14th European Biomass Conference & Exhibition. October 2005, Paris.

BIOMASS Energy Centre. 2013. Typical calorific values of fuels. [Online]. Available from: http://www.biomassenergycentre.org.uk/portal/page?_pageid=75,20041&_dad=portal&_schema=PORTAL [Accessed: 13 May 2013].

BLRB. 2012. Bayannur overall plan on landuse 2006–2020. [Online]. 2012. Available from: http://www.nmggtt.gov.cn/nmgt/html/bayannaoer/col5010/2012-03/30/20120330113456177973861_1.html [Accessed: 20 November 2012].

Boeuf-Tremblay, V., Plantureux, S. & Guckert, A. 2005. Influence of mechanical impedance on root exudation of maize seedlings at two development stages. Plant and Soil, 172: 279–287.

Bonanno, G. & Lo Giudice, R. 2010. Heavy metal bioaccumulation by the organs of *Phragmites australis* (common reed) and their potential use as contamination indicators. Ecological Indicators, 10(3): 639–645.

Boothroyd, I.K.G. & Stark, J. 2000. Use of invertebrates in monitoring. In: K. J. Collier & M. J. Winterbourn eds. New Zealand stream invertebrates: ecology and implications for management. Christchurch: New Zealand Limnological Society, NIWA, 344–373.

Borin, S., Brusetti, L., Mapelli, F., D'Auria, G., Brusa, T., Marzorati, M., Rizzi, A., Yakimov, M., Marty, D., De Lange, G.J., Van der Wielen, P., Bolhuis, H., McGenity, T.J., Polymenakou, P.N., Malinverno, E., Giuliano, L., Corselli, C. & Daffonchio, D. 2009. Sulfur cycling and methanogenesis primarily drive microbial colonization of the highly sulfidic Urania deep hypersaline basin. Proceedings of the National Academy of Sciences of the United States of America, 106(23): 9151–9156.

Borruso, L. 2014. Rhizobacterial communities as bioindicators of environmental stresses in freshwater ecosystems. PhD Thesis. Bolzano, Italy: Free University of Bozen-Bolzano.

Bradshaw, A.D. & Chadwick, M.J. 1980. The restoration of land. University of California Press, Berkeley, Calif.

Bragato, C., Brix, H. & Malagoli, M. 2006. Accumulation of nutrients and heavy metals in *Phragmites australis* (Cav.) Trin. ex Steudel and *Bolboschoenus maritimus* (L.) Palla in a constructed wetland of the Venice lagoon watershed. Environmental Pollution, 144(3): 967–975.

Breckle, S.W., Wucherer, W., Dimeyeva, L.A. & Ogar, N.P. 2012. Aralkum – a Man-Made Desert. Springer, Heidelberg.

Bridgeman, T.G., Jones, J.M., Shield, I. & Williams, P.T. 2008. Torrefaction of reed canary grass, wheat straw and willow to enhance solid fuel qualities and combustion properties. Fuel, 87(6): 844–856.

Bridgwater, A. V. 2004. Biomass fast pyrolysis. Thermal Science, 8(2): 21–49.

Brown, A. & Imberger, J. 2006. The index of sustainable functionality (ISF): A prospective tool for assessing the sustainability of the impact of the World Bank projects. Report. Perth, Western Australia: Centre for Water Research.

Bruelheide, H., Manegold, M. & Jandt, U. 2004. The genetical structure of *Populus euphratica* and *Alhagi sparsifolia* stands in the Taklamakan Desert. In: M. Runge & X. M. Zhang eds. Ecophysiology and habitat requirements of perennial plant species in the Taklamakan Desert. Aachen: Shaker, 1–153.

Brundtland, G.H. 1987. Our Common Future: The World Commission on Environment and Development. Oxford: Oxford University Press.

Cardinale, M., Brusetti, L., Quatrini, P., Borin, S., Puglia, A.M., Rizzi, A., Zanardini, E., Sorlini, C., Corselli, C. & Daffonchio, D. 2004. Comparison of different primer sets for use in automated ribosomal intergenic spacer analysis of complex bacterial communities. Applied and Environmental Microbiology, 70(10): 6147–6156.

Chambers, R.M., Meyerson, L.A. & Saltonstall, K. 1999. Expansion of *Phragmites australis* into tidal wetlands of North America. Aquatic Botany, 3–4: 261–273.

Chang-wei, L., Jiang, H., Ying, L., Hai-fang, M., Hua-lin, L. & Feng-jiao, W. 2010. Acta Ecologica Sinica Temporal and spatial distribution of biogenic silica and its significance in the Wuliangsuhai Lake and Daihai Lake. Acta Ecologica Sinica, 30(2): 100–105.

Chapin, F.S., Matson, P.A. & Mooney, H.A. 2002. Principles of Terrestrial Ecosystem Ecology. 2nd Ed. F Stuart Chapin III, P. A. Matson, & H. A. Mooney eds. New York: Springer-Verlag.

Chen, H., Jia, B. & Lau, S.S.Y. 2008. Sustainable urban form for Chinese compact cities: Challenges of a rapid urbanized economy. Habitat International, 32(1): 28–40.

Chen, H., Zhu, Q., Peng, C., Wu, N., Wang, Y., Fang, X., Jiang, H., Xiang, W., Chang, J., Deng, X. & Yu, G. 2013. Methane emissions from rice paddies natural wetlands, lakes in China: Synthesis new estimate. Global Change Biology, 19(1): 19–32.

Cheng, S. 2003. Heavy metal pollution in China: Origin, pattern and control. Environmental Science and Pollution Research International, 10(3): 192–198.

Chivu, I.A. 1968. The use of reeds as raw material for pulp and paper production. In: FAO ed. Practical experiment in the cropping of reeds for the manufacture of pulp and paper: Economic results. Food and Agriculture Organization of the United Nations, 877–899.

Cirella, G.T. & Tao, L. 2008. Measuring sustainability: an application using the index of sustainable functionality in South East Queensland, Australia. The International Journal of Interdisciplinary Social Sciences, 3(8): 231–240.

Cirella, G.T. & Zerbe, S. 2014a. Index of sustainable functionality: Procedural developments and application in Urat Front Banner, Inner Mongolia Autonomous Region. The International Journal of Environmental Sustainability, In press.

Cirella, G.T. & Zerbe, S. 2014b. Quizzical societies: A closer look at sustainability and principles of unlocking its measurability. The International Journal of Science in Society, 5(3): 29–45.

CMF. 1990. The Forests of Xinjiang. Urumqi: China Ministry of Forestry, Xinjiang Peoples Publishing House.

Cocking, E.C. 2003. Endophytic colonization of plant roots by nitrogen-fixing bacteria. Plant and Soil, 252(1): 169–175.

Compant, S., Reiter, B., Sessitsch, A., Nowak, J., Clément, C. & Ait Barka, E. 2005. Endophytic colonization of Vitis vinifera L. by plant growth-promoting bacterium Burkholderia sp. strain PsJN. Applied and Environmental Microbiology, 71(4): 1685–1693.

Conedera, M., Krebs, P., Tinner, W., Pradella, M. & Torriani, D. 2004. The cultivation of *Castanea sativa* (Mill.) in Europe, from its origin to its diffusion on a continental scale. Vegetation History and Archaeobotany, 13: 161–179.

Conkle, J.L., White, J.R. & Metcalfe, C.D. 2008. Reduction of pharmaceutically active compounds by a lagoon wetland wastewater treatment system in Southeast Louisiana. Chemosphere, 73(11): 1741–1748.

Craine, J.M., Lee, W.G., Bond, W.J., Williams, R.J. & Johnson, L.C. 2005. Environmental constraints on a global relationship among leaf and root traits of grasses. Ecology, 86(1): 12–19.

Cubars, E. 2010. Analysis of reed-covered surface and of potential biomass in Latvia's lakes. COFREEN: European Regional Development Fund.

Curtis, T.P., Sloan, W.T. & Scannell, J.W. 2002. Estimating prokaryotic diversity and its limits. Proceedings of the National Academy of Sciences of the United States of America, 99(16): 10494–10499.

Daily, G.C. 1995. Restoring Value to the World's Degraded Lands. Science, 269(5222): 350–354.

Daly, H.E. 1996. Beyond Growth: The Economics of Sustainable Development. Boston: Beacon Press.

Daly, H.E. 2006. The Future of Sustainability. M. Keiner ed. Dordrecht, The Netherlands: Springer.

Daly, H.E. & Cobb, J. 1989. For the common good: redirecting the economy toward community, the environment, and a sustainable future. Boston: Beacon Press.

Davies, G.R. 2013. Appraising Weak and Strong Sustainability: Searching for a Middle Ground. Consilience: The Journal of Sustainable Development, 10(1): 111–124.

Dela Cruz, A. 1987. The Production of pulp from marsh grass. Economic Botany, 32: 46–50.

Devuyst, D. 1999. Sustainability assessment: The application of a methodological framework. Journal of Environmental Assessment Policy and Management, 01(04): 459–487.

Dienst, M., Schmieder, K., & Ostendorp, W. (2004). Dynamik der Schilfröhrichte am Bodensee unter dem Einfluss von Wasserstandsvariationen. Limnologica - Ecology and Management of Inland Waters, 34(1–2): 29–36.

Döring, R. 2009. Natural capital – What's the difference? In: R. Döring ed. Sustainability, natural capital and nature conservation. Marburg: Metropolis-Verlag, 123–142.

Dou, J., Liu, X. & Ding, A. 2009. Anaerobic degradation of naphthalene by the mixed bacteria under nitrate reducing conditions. Journal of Hazardous Materials, 165(1–3): 325–331.

Dowling, M. & Wignaraja, G. 2006. Central Asia after fifteen years of transition: Growth, regional cooperation, and policy choices. Asia-Pacific Development Journal, 13(2): 113–144.

EC2. 2011. Study of potential and constraints of the biomass sector in China - Executive Summary. [Online]. 2011. Europe-China Clean Energy Centre. Available from: http://www.ec2.org.cn/ [Accessed: 10 September 2013].

Economist Intelligence Unit. 2005. Quality-of-life Index. [Online]. 2005. Available from: http://www.eiu.com/ [Accessed: 20 December 2013].

Egamberdieva, D., Kamilova, F., Validov, S., Gafurova, L., Kucharova, Z. & Lugtenberg, B. 2008. High incidence of plant growth-stimulating bacteria associated with the rhizosphere of wheat grown on salinated soil in Uzbekistan. Environmental Microbiology, 10(1): 1–9.

Eissenstat, D.M. & Yanai, R.D. 1997. The Ecology of Root Lifespan. Advances in Ecological Research, 27(C): 1–60.

Ejina Qizhe. 1998. Ejina Qizhe (Description of Ejina County). Beijing: Fangzhe Chubanshe.

Elkington, J. 1998. Cannibals With Forks: The Triple Bottom Line of 21st Century Business. Stony Creek, CT: New Society Publishers.

Emmanuel, E., Keck, G., Blanchard, J.M., Vermande, P. & Perrodin, Y. 2004. Toxicological effects of disinfections using sodium hypochlorite on aquatic organisms and its contribution to AOX formation in hospital wastewater. Environment International, 30(7): 891–900.

Esty, D.C., Levy, M., Srebotnjak, T. & de Sherbinin, A. 2005. 2005 Environmental Sustainability Index Benchmarking National Environmental Stewardship. New Haven: Yale Center for Environmental Law & Policy.

Eusemann, P., Petzold, A., Thevs, N. & Schnittler, M. 2013. Growth patterns and genetic structure of *Populus euphratica* Oliv. (Salicaceae) forests in NW China - Implications for conservation and management. Forest Ecology and Management, 297: 27–36.

Faafeng, B., Li, T., Lindblom, E., Ye, J., Oredalen, T.J., Lövik, J.E.L. & Svenson, A. 2008. Lake Wuliangsuhai Restoration Project: Water Quality Monitoring System. Norwegian Agency for Development Cooperation Agency.

Fachagentur für nachwachsende Rohstoffe. 2012. Nachwachsende Rohstoffe in der Industrie (Renewable raw material for industry). BioCat Project Group.

Fang, J., Wang, Z., Zhao, S., Li, Y., Tang, Z., Yu, D., Ni, L., Liu, H., Xie, P., Da, L., Li, Z. & Zheng, C. 2006. Biodiversity changes in the lakes of the Central Yangtze. Frontiers in Ecology and the Environment, 4(7): 369–377.

FAO. 2011. The State of the World's Land and Water Resources for Food and Agriculture (SOLAW): Managing systems at risk. London: Earthscan Publications Ltd.

Fejes, J., Ratnaweera, H., Yawei, L., Lindblim, E. & Faafeng, B. 2008. Inner Mongolia Lake Restoration Project, Lake Wuliangsuhai Comprehensive Study Extension, Final Report. Norwegian Institute for Water Research.

Feng, Q. & Cheng, G.D. 1998. Current situation, problem and rational utilisation of water resources in Gansu Province. Chinese J. Arid Land Research, 11: 293–299.

Feng, X.Q., Zhang, G.X. & Jun Xu, Y. 2013. Simulation of hydrological processes in the Zhalong wetland within a river basin, Northeast China. Hydrology and Earth System Sciences, 17(7): 2797–2807.

Fisher, M.M. & Triplett, E.W. 1999. Automated approach for ribosomal intergenic spacer analysis of microbial diversity and its application to freshwater bacterial communities. Applied and Environmental Microbiology, 65(10): 4630–4636.

Foliente, G., Kearns, A., Maheepala, S., Bai, X. & Barnett, G. 2007. Beyond Triple Bottom Line – Sustainable Cities: CSIRO. In: State of Australian Cities National Conference (SOAC2007). 2007, Adelaide, Australia: Commonwealth Scientific and Industrial Research Organisation, 28–30 November.

Franz, A., Burgstaller, W. & Schinner, F. 1991. Leaching with *Penicillium simplicissimum*: Influence of Metals and Buffers on Proton Extrusion and Citric Acid Production. Applied and Environmental Microbiology, 57(3): 769–774.

Fränzle, O. 2006. Complex bioindication and environmental stress assessment. Ecological Indicators, 6(1): 114–136.

Freschet, G.T., Cornelissen, J.H.C., van Logtestijn, R.S.P. & Aerts, R. 2010. Evidence of the `plant economics spectrum' in a subarctic flora. Journal of Ecology, 98(2): 362–373.

Friberg, R. & Blasiak, W. 2002. Measurements of mass flux and stoichiometry of conversion gas from three different wood fuels as function of volume flux of primary air in packed-bed combustion. Biomass and Bioenergy, 23(3): 189–208.

Frick, A., Steffenhagen, P., Zerbe, S., Timmermann, T. & Schulz, K. 2011. Monitoring of the vegetation composition in rewetted peatland with iterative decision tree classification of satellite imagery. Photogrammetrie - Fernerkundung - Geoinformation, 2011(3): 109–122.

Fu, G.B., Chen, S.L., Liu, C.M. & Shepard, D. 2004. Hydro-climatic trends of the Yellow River basin for the last 50 years. Climatic Change, 65(1–2): 149–178.

Gahlert, F. 2006. Schilf (*Phragmites australis*) in der Tarim-Aue, Nordwest China. [Online]. 2006. Available from: http://laoek.botanik.uni-greifswald.de/literatur/diplom/GAHLERT [Accessed: 1 September 2013].

Gamauf, N. 2000. Satellitenbildauswertung des Schilfgürtels am Neusiedler See zur Ermittlung von Rohstoffpotenzialen. PhD Thesis. Vienna, Austria: University of Natural Resources and Life Sciences.

Garcia-Perez, M. & Metcalf, J. 2008. The Formation of Polyaromatic Hydrocarbons and Dioxins During Pyrolysis: A Review of the Literature with Descriptions of Biomass Composition, Fast Pyrolysis Technologies and Thermochemical Reactions. Pullman: Washington State University.

Gibson, R., Agnolin, J., Hassan, S., Lawrence, D., Robinson, J.B., Tansey, J., Watson, J. & Whitelaw, G.S. 2001. Specification of sustainability-based environmental assessment decision criteria and implications for determining significance in environmental assessment. [Online]. 2001. Available from: http://www.sustreport.org/ [Accessed: 20 December 2013].

Giese, E., Bahro, G. & Betke, D. 1998. Umweltzerstoerungen in Trockengebieten Zentralasiens (West- und Ost-Turkestan): Ursachen, Auswirkungen, Maßnahmen. Stuttgart: Steiner.

Giese, E. & Moßig, I. 2004. Klimawandel in Zentralasien. Zentrum für internationale Entwicklungs- und Umweltforschung (ZEU). Giessen: Justus-Liebig-Universität Giessen.

Girotti, S., Ferri, E.N., Fumo, M.G. & Maiolini, E. 2008. Monitoring of environmental pollutants by bioluminescent bacteria. Analytica Chimica Acta, 608(1): 2–29.

Glantz, M.H. 2005. Water, Climate, and Development Issues in the Amu Darya Basin. Mitigation and Adaptation Strategies for Global Change, 10(1): 23–50.

Glick, B.R. 2012. Plant Growth-Promoting Bacteria: Mechanisms and Applications. Scientifica, 2012: 1–15.

Glick, B.R. 2010. Using soil bacteria to facilitate phytoremediation. Biotechnology Advances, 28(3): 367–374.

Gomez, C. & Bosecker, K. 1999. Leaching heavy metals from contaminated soil by using *Thiobacillus ferrooxidans* or *Thiobacillus thiooxidans*. Geomicrobiology Journal, 16(3): 233–244.

Goodwin, D. 2011. Cantera: An object-oriented software toolkit for chemical kinetics, thermodynamics, and transport processes. [Online]. 2011. Available from: http://cantera.github.io/docs/sphinx/html/index.html# [Accessed: 1 December 2011].

Gordon, T.J. 1994. The Delphi Method. In: AC/UNU Millennium Project, Future Research Methodology.

Graneli, W. 1984. Reed *phragmites australis* (Cav.) Trin. ex Steudel as an energy source in Sweden. Biomass, 4(3): 183–208.

Graneli, W. 1990. Standing crop and mineral content of reed, *Phragmites australis,* in Sweden - Management of reed stands to maximize harvestable biomass. Folia Geobotanica, 25(3): 291–302.

Gries, D., Zeng, F., Foetzki, A., Arndt, S.K., Bruelheide, H., Thomas, F.M., Zhang, X. & Runge, M. 2003. Growth and water relations of *Tamarix ramosissima* and *Populus euphratica* on Taklamakan desert dunes in relation to depth to a permanent water table. Plant, Cell and Environment, 26: 725–736.

Grime, J.P. 1977. Evidence for the Existence of Three Primary Strategies in Plants and Its Relevance to Ecological and Evolutionary Theory. The American Naturalist, 111(982): 1169.

Gruschke, A. 1991. Neulanderschließung in Trockengebieten der Volksrepublik China. Hamburg, Germany: Institut für Asienkunde.

Guo, Q., Feng, Q. & Li, J. 2009. Environmental changes after ecological water conveyance in the lower reaches of Heihe River, northwest China. Environmental Geology, 58(7): 1387–1396.

Habermas, J. 1965. Erkenntnis und Interesse. In: A. Hans & E. Topitsch eds. Werturteilsstreit, Darmstadt. Wissenschaftliche Buch Gesellschaft, 1990, 334–352.

Häkkinen, J. 2007. Traditional use of reed. In: Read Up on Reed! 62–72.

Hale, M.G., Foy, C.L. & Shay, F.J. 1971. Factors affecting root exudation. Advances in Agronomy, 23: 89–109.

Hall, R.M. & Collis, C.M. 1995. Mobile gene cassettes and integrons: capture and spread of genes by site-specific recombination. Molecular Microbiology, 15(4): 593–600.

Hampicke, U. 2009. Kosten der Renaturierung. In: S. Zerbe & G. Wiegleb eds. Renaturierung von Ökosystemen in Mitteleuropa. Spektrum Akademischer Verlag, Heidelberg, 441–457.

Han, W.X., Fang, J.Y., Reich, P.B., Ian Woodward, F. & Wang, Z.H. 2011. Biogeography and variability of eleven mineral elements in plant leaves across gradients of climate, soil and plant functional type in China. Ecology Letters, 14(8): 788–796.

Hansmann, P. 2008a. "They Call Them Golden Sticks"- Socio-economic Explorations Around the Commodity of Reed. Wageningen University, The Netherlands.

Hansmann, P. 2008b. "They call them golden sticks"- Socio-economic explorations around the commodity of reed.

Hansson, P.-A. & Fredriksson, H. 2004. Use of summer harvested common reed (*Phragmites australis*) as nutrient source for organic crop production in Sweden. Agriculture, Ecosystems & Environment, 102(3): 365–375.

Haslam, S.M. 2010. A book of reed: (*Phragmites australis* (Cav.) Trin. ex Steudel, formerly *Phragmites communis* Trin.). Forrest Text.

Hawke, C.J. & Jose, P. V. 1996. Reedbed management for commercial and wildlife interests. RSPB.

He, J.-S., Wang, L., Flynn, D.F.B., Wang, X., Ma, W. & Fang, J. 2008. Leaf nitrogen: Phosphorus stoichiometry across Chinese grassland biomes. Oecologia, 155(2): 301–310.

He, Z.B. & Zhao, W.Z. 2006. Characterising the spatial structures of riparian plant communities in the lower reaches of the Heihe River in China using geostatistical techniques. Ecological Research, 21: 551–559.

Hedin, S.A. 1943. Reports from the Scientific Expedition to the North-Western Provinces of China under the Leadership of Dr. Sven Hedin. Part I. Stockholm.

Heuer, H. & Smalla, K. 2007. Horizontal gene transfer between bacteria. Environmental Biosafety Research, 6(1–2): 3–13.

Hijosa-Valsero, M., Fink, G., Schlüsener, M.P., Sidrach-Cardona, R., Martín-Villacorta, J., Ternes, T. & Bécares, E. 2011. Removal of antibiotics from urban wastewater by constructed wetland optimization. Chemosphere, 83(5): 713–719.

Holzmann, G. & Wangelin, M. 2009. printed. Natürliche und pflanzliche Baustoffe: Rohstoff - Bauphysik - Konstruktion [Gebundene Ausgabe] (Natural and plant construction materials: resources construction physik-Construction). Vieweg and Teubner.

Hoppe, T. 1992. Chinesische Agrarpolitik und Uygurische Agrarkultur im Widerstreit. Das sozio-kulturelle Umfeld von Bodenversalzungen und -alkalisierungen im nördlichen Tarim-Becken (Xinjiang). Hamburg, Germany: Institut für Asienkunde.

Hou, L.G., Xiao, H.L., Si, J.H., Xiao, S.C., Zhou, M.X. & Yang, Y.G. 2010. Evapotranspiration and crop coefficient of *Populus euphratica* Oliv forest during the growing season in the extreme arid region northwest China. Agricultural Water Management, 97(2): 351–356.

Hou, P., Beeton, R.J.S., Carter, R.W., Dong, X.G. & Li, X. 2007. Response to environmental flows in the Lower Tarim River, Xinjiang, China: An ecological interpretation of water-table dynamics. Journal of Environmental Management, 83(4): 383–391.

Hu, J. & Ge, Y. 2004. Water allocation and coordination systems in the Yellow River (huanghe shuiziyuan de fenpei moshi yu xietiao jizhi). Management World (guanli shijie), 8(66): 49–58.

Hu, J. 2012. Firmly march on the path of socialism with chinese charcaterstics and strive to complete the building of a moderately prosperous society in all respects. [Online]. 2012. Report to the Eighteenth National Congress of the Communist Party of China on 8 November 2012. Available from: http://www.chinadaily.com.cn/china/2012cpc/2012-11/18/content_15939493.htm [Accessed: 30 April 2014].

Hu, R., Yang, C., Ma, H., Jiang, F. & Urkunbek, A. 1994. Glaciers and Lakes in the Tianshan Mountains and Climate Trends. Arid Land Geography (Ganhan Qu Dili), 17: 1–9.

Huang, P.Y. 1986. A Preliminary Study on the Decline of the Distribution Range and Regeneration of the Forest Land of Populus euphratica in the Tarimpendi (Basin). Acta Phytoecologica et Geobotanica Sinica, 10: 302–309.

Hubacek, K., Guan, D., Barrett, J. & Wiedmann, T. 2009. Environmental implications of urbanization and lifestyle change in China: Ecological and Water Footprints. Journal of Cleaner Production, 17(14): 1241–1248.

Iital, A., Klõga, M., Kask, Ü., Voronova, V. & Cahill, B. 2012. Reed harvesting. In: A. Schultz-Zehden & M. Matczak eds. Compendium. An Assessment of Innovative and Sustsainable Uses of Baltic Marine Resources. 103–124.

Imberger, J., Mamouni, E.D., Anderson, J., Ng, M., Nicol, S. & Veale, A. 2007. The index of sustainable functionality: a new adaptive, multicriteria measurement of sustainability – application to Western Australia. International Journal of Environment and Sustainable Development, 6(3): 323–355.

Institute for Economics and Peace. 2007. Global Peace Index. [Online]. 2007. Available from: http://economicsandpeace.org/ [Accessed: 12 January 2014].

IUCN. 2008. International Union for Conservation of Nature Website. [Online]. 2008. Available from: http://www.iucn.org/ [Accessed: 20 December 2013].

Jax, K., Barton, D.N., Chan, K.M.A., de Groot, R., Doyle, U., Eser, U., Görg, C., Gòmez-Baggethun, E., Griewald, Y., Haber, W., Haines-Young, R., Heink, U., Jahn, T., Joosten, H., Kerschbaumer, L., Korn, H., Luck, G.W., Matzdorf, B., Muraca, B., Neßhöver, C., Norton, B., Ott, K., Potschin, M., Rauschmayer, F., von Haaren, C. & Wichmann, S. 2013. Ecosystem services and ethics. Ecological Economics, 93: 260–268.

Ji, Y.H., Zhou, G.S., Lv, G.H., Zhao, X.L. & Jia, Q.Y. 2009. Expansion of *Phragmites australis* in the Liaohe Delta, north-east China. Weed Research, 49(6): 613–620.

Jiang, F., Ma, H., Hu, R. & Yuan, Y. 1997. Response of Water Resources in Xinjiang to Future Climate Changes in Central Asia. Arid Land Geography (Ganhan Qu Dili), 20: 40–46.

Jiang, X. & Liu, C. 2010. The influence of water regulation on vegetation in the lower Heihe River. Journal of Geographical Sciences, 20(5): 701–711.

Jin, X., Schaepman, M., Clevers, J., Su, Z. & Hu, G. 2010. Correlation between annual runoff in the Heihe River to the vegetation cover in the Ejina Oasis (China). Arid Land Research and Management, 24(1): 31–41.

Jolivet, M., Barrier, L., Dominguez, S., Guerit, L., Heilbronn, G. & Fu, B. 2014. Unbalanced sediment budgets in the catchment–alluvial fan system of the Kuitun River (northern Tian Shan, China): Implications for mass-balance estimates, denudation and sedimentation rates in orogenic systems. Geomorphology, 214: 168–182.

Jordan, C., Ayres, T.R., Brown, A., Linsky, J.L. & Simon, T. 1987. Monthly Notices of the Royal Astronomical Society. Oxford University Press.

Junfeng, L., Runqing, H., Yanqin, S., Jingli, S., Bhattacharya, S.C. & Abdul Salam, P. 2005. Assessment of sustainable energy potential of non-plantation biomass resources in China. Biomass and Bioenergy, 29(3): 167–177.

Kaltschmitt, M., Streicher, W. & Wiese, A. 2007. Renewable Energy: Technology, Economics and Environment. Berlin: Springer Berlin Heidelberg.

Kask, Ü., Kask, L. & Paist, A. 2007. Reed as energy resource in Estonia. In: I. Ikonen & E. Hagelberg eds. Read Up on Reed! Turku: Vammalan Kirjapaino Oy, 102–114.

Kask, Ü. 2011. Reed as bio-energy: opportunities to use it in boiler-houses and as biogas source. In: Reed for bio-energy and Construction. 2011, 11 March 2011: Kaarina, Finland.

Kennedy, I.R., Pereg-gerk, L.L., Wood, C., Deaker, R. & Gilchrist, K. 1997. Biological nitrogen fixation in non-leguminous field crops: Facilitating the evolution of an effective association between Azospirillum and wheat. Plant and Soil, 194(1-2): 65–79.

Kentula, M.E. 2002. Wetland Restoration and Creation. [Online]. 2002. Wetlands Research Program, U.S. Environmental Protection Agency. Available from: https://water.usgs.gov/nwsum/WSP2425/restoration.html [Accessed: 4 May 2014].

Kerkhoff, A.J., Fagan, W.F., Elser, J.J. & Enquist, B.J. 2006. Phylogenetic and growth form variation in the scaling of nitrogen and phosphorus in the seed plants. The American Naturalist, 168(4): E103–E122.

Kerschbaumer, L., Köbbing, J.F., Ott, K., Thevs, N. & Zerbe, S. 2014. Development Scenarios on Hetao Irrigation Area (China) – A qualitative analysis from social, economic and ecological perspectives. Environmental Earth Sciences, DOI: 10.1007/s12665–014–3061–8.

Kerschbaumer, L. & Ott, K. 2013. Maintaining a River's Healthy Life? An Inquiry on Water Ethics and Water Praxis in the Upstream Region of China's Yellow River. Water Alternatives, 6(1): 107–124.

Kim, J.S.S. 2007. A China Environmental Health Project Fact Sheet: Transboundary Air Pollution—Will China Choke On Its Success? [Online]. 2007. Available from: http://www.wilsoncenter.org/sites/default/files/transboundary_feb2.pdf [Accessed: 14 March 2014].

Kim, K.Y., Jordan, D. & McDonald, G.A. 1998. Enterobacter agglomerans, phosphate solubilizing bacteria, and microbial activity in soil: Effect of carbon sources. Soil Biology and Biochemistry, 30(8–9): 995–1003.

Kim, Y.C., Leveau, J., McSpadden Gardener, B.B., Pierson, E.A., Pierson, L.S. & Ryu, C.-M. 2011. The multifactorial basis for plant health promotion by plant-associated bacteria. Applied and Environmental Microbiology, 77(5): 1548–1555.

Kitzler, H., Pfeifer, C. & Hofbauer, H. 2012. Combustion of Reeds in a 3 MW District Heating Plant. International Journal of Environmental Science and Development, 3(4): 407–411.

Kitzler, H., Pfeifer, C. & Hofbauer, H. 2011. Gasification of reed in a 100kW dual fluidized bed steam gasification. In: 19th European Biomass Conference & Exhibition. June 2011, Berlin, Germany, 1101–1105.

Kloepper, J.W., Gutiérrez-Estrada, A. & McInroy, J.A. 2007. Photoperiod regulates elicitation of growth promotion but not induced resistance by plant growth-promoting rhizobacteria. Canadian Journal of Microbiology, 53(2): 159–167.

Knoef, H.A.M. 2005. Practical aspects of biomass gasification. In: H. A. M. Knoef ed. Handbook Biomass Gasification. Meppel: Biomass Technology Group BV.

Köbbing, J.F., Beckmann, V., Thevs, N., Peng, H. & Zerbe, S. 2014a. Investigation of a reed economy (*Phragmites australis*) under threat: pulp and paper market, values and netchain at Wuliangsuhai Lake, Inner Mongolia, China. Technical Report. Greifswald, Germany: Institute for Botany and Landscape Ecology, University Greifswald.

Köbbing, J.F., Patuzzi, F., Baratieri, M., Beckmann, V., Thevs, N., & Zerbe, S. 2014b. Economic evaluation of common reed potential for energy production: A case study in Wuliangsuhai Lake (Inner Mongolia, China). Biomass and Bioenergy, 70, 315–329. doi:10.1016/j.biombioe.2014.08.002

Köbbing, J.F., Thevs, N. & Zerbe. 2013. The utilisation of Reed (*Phragmites australis*) – A review. Mires and Peat, 13: 1–14.

Kole, M.M., Page, W.J. & Altosaar, I. 1988. Distribution of Azotobacter in Eastern Canadian soils and in association with plant rhizospheres. Canadian Journal of Microbiology, 34: 815–817.

Komulainen, M., Simi, P., Hagelberg, E., Ikonen, I. & Lyytinen, S. 2008. Reed energy - Possibilities of using the common reed for energy generation in southern Finland. Turku: Turku University of Applied Sciences.

Koziński, J.A., Saade, R. & Zheng, G. 1996. Transformations of sludge waste during combustion in a low-high-low temperature reactor. Symposium (International) on Combustion, 26(2): 2495–2502.

Krebs, A. 1999. Ethics of Nature. Berlin: DeGruyter Verlag.

Kreuzberg, E. 2005. Ecosystem approach in basin management in Central Asia: From theory to practice (on the example of Ili-Balkhash Basin). In: International Meeting on the Implementation of the European Water Framework Directive, 29 September – 1 October 2005. 2005, Namur, Belgium: Regional Environmental Centre of Central Asia (CAREC).

Kristiana, R. 2009. Sustainability assessment in water management: The application and development of the Index of Sustainable Functionality on various water resources. Perth, Western Australia: University of Western Australia.

Krivitzki, A.I. 1959. Das Schilfrohr als Rohstoff für die Erzeugung von Baumaterialien - Primen. Kam. V. Stroit. Moscow: Gosstroiizdat.

Kronbergs, E., Kaktis, A., Smits, M. & Nulle, I. 2006. Biomass condition for solid biofuel compositions. In: M. Barz & M. Alhhaus eds. Use of Bioenergy in the Baltic Sea Region. 2006, 83–92.

Kronbergs, E. & Kronbergs, A. 2011. Common reed as solid biofuel resource. In: 8–9 September 2011 ed. COFREEN Project. 2011, Matsalu, Estonia: European Regional Development Fund.

Ku, C.S. & Mun, S.P. 2006. Characterization of Pyrolysis Tar Derived from Lignocellulosic Biomass. Journal of Industrial and Engineering Chemistry, 12(6): 853–861.

Kuhlman, T., Diogo, V. & Koomen, E. 2013. Exploring the potential of reed as a bioenergy crop in the Netherlands. Biomass and Bioenergy, 55: 41–52.

Kumar, M. & Khanna, S. 2010. Diversity of 16S rRNA and dioxygenase genes detected in coal-tar-contaminated site undergoing active bioremediation. Journal of Applied Microbiology, 108(4): 1252–1262.

Kusler & Kentula. 1990. Wetland creation and restoration: The status of the science. 2nd Ed. Washington, D.C.: Island Press.

Kuzmina, Z. V & Treshkin, S.Y. 1997. Soil salinization and dynamics of Tugai vegetation in the southwestern Caspian Sea region and in the Aral Sea coastal region. Eurasien Soil Science, 30: 642–649.

Kwapinski, W., Byrne, C.M.P., Kryachko, E., Wolfram, P., Adley, C., Leahy, J.J., Novotny, E.H. & Hayes, M.H.B. 2010. Biochar from Biomass and Waste. Waste and Biomass Valorization, 1(2): 177–189.

Lambertini, C., Gustafsson, M.H.G., Frydenberg, J., Speranza, M. & Brix, H. 2008. Genetic diversity patterns in *Phragmites australis* at the population , regional and continental scales. Aquatic Botany, 88: 160–170.

Lang, C. 2007. Pulp Mill Watch Factsheet. Country profiles: China. [Online]. 2007. Available from: www.pulpmillwatch.org [Accessed: 15 February 2014].

Lange, T. 2006. Orientierende Versuche zum Einsatz von Getreide und halmgutartiger Biomasse in einer Kleinfeuerungsanlage. FH Stralsund.

Lauber, C.L., Strickland, M.S., Bradford, M.A. & Fierer, N. 2008. The influence of soil properties on the structure of bacterial and fungal communities across land-use types. Soil Biology and Biochemistry, 40(9): 2407–2415.

Lavrenko, E.M. 1956. Karta rastitelnosti srednej Asii (Vegetation Map of Central Asia). Moscow: Akademijca a NAUK SSSR.

Lear, G. & Lewis, G.D. 2009. Impact of catchment land use on bacterial communities within stream biofilms. Ecological Indicators, 9(5): 848–855.

Lee, H. 2011. Handbook of Bioenergy Crops. A Complete Reference to Species, Development and Applications. International Journal of Agricultural Sustainability, 9(3): 473.

Leonard, K.M. & Swanson, G.W. 2001. Comparison of operational design criteria for subsurface flow constructed wetlands for wastewater treatment. Water Science and Technology, 43: 301–307.

Leung, D.Y.C., Yin, X.L. & Wu, C.Z. 2004. A review on the development and commercialization of biomass gasification technologies in China. Renewable and Sustainable Energy Reviews, 8(6): 565–580.

Li, C. & Suzuki, K. 2009. Tar property, analysis, reforming mechanism and model for biomass gasification—An overview. Renewable and Sustainable Energy Reviews, 13(3): 594–604.

Li, L., Han, W., Thevs, N., Jia, X., Ji, C., Jin, D., He, P., Schmitt, A.O., Cirella, G.T. & Zerbe, S. 2014a. A Comparison of the Functional Traits of Common Reed (Phragmites australis) in Northern China: Aquatic vs. Terrestrial Ecotypes. PloS ONE, 9(2): e89063. doi:10.1371/journal.pone.0089063

Li, L., Zerbe, S., Han, W., Thevs, N., Li, W., He, P., Schmitt, A.O., Liu, Y. & Ji, C. 2014b. Nitrogen and phosphorus stoichiometry of common reed (*Phragmites australis*) and its relationship to nutrient availability in northern China. Aquatic Botany, 112: 84–90.

Li, P., Yang, T., Shi, H., Wu, X. & Qin, Y. 2012a. Income and expenditure of the herdsman in Inner Mongolia Autonomous Region: A case study of

Xianghuang Banner in Xilin Gol League. Asian Agricultural Research, 4(9): 26–39.

Li, Q., Feng, Q. & Zhai, L. 2010a. Study of the height growth dynamic based on tree-ring data in *Populus euphratica* from the lower reach of the Heihe River, China. Dendrochronologia, 28(1): 49–64.

Li, X., Lua, L., Yangc, W. & Chenga, G. 2012b. Estimation of evapotranspiration in an arid region by remote sensing-A case study in the middle reaches of the Heihe River Basin. International Journal of Applied Earth Observation and Geoinformation, 17(1): 85–93.

Li, Y. & Zhang, M. 2012. "The Pearl of the Northern Frontier" Wuliangsuhai may disappear. [Online]. 2012. Economic Information. Available from: http://dz.jjckb.cn/www/pages/webpage2009/html/2012-08/06/content_48241.htm?div=-1 [Accessed: 20 December 2012].

Li, Y.H., Zhu, J.N., Zhai, Z.H. & Zhang, Q. 2010b. Endophytic bacterial diversity in roots of *Phragmites australis* in constructed Beijing Cuihu Wetland (China). FEMS Microbiology Letters, 309: 84–93.

Liao, Z. 1993. The environmental chemistry and biological effects of microelement. Beijing: China Environmental Science Press.

Link, S., Kask, Ü., Paist, A., Siirde, A., Arvelakis, S., Hupa, M., Yrjas, P. & Külaots, I. 2013. Reed as a gasification fuel: a comparison with woody fuels. Mires and Peat, 13: 1–12.

Liu, H., Jiang, G., Zhuang, H. & Wang, K. 2008. Distribution, utilization structure and potential of biomass resources in rural China: With special references of crop residues. Renewable and Sustainable Energy Reviews, 12(5): 1402–1418.

Liu, M.G. 1997. Atlas of Physical Geography of China. Beijing: China Map Press.

Liu, P.J., Zhang, L. & Fan, C.Q. 1990. The *Populus euphratica* resources along the Tarim River . Science, Technique,. K. Liang & P. J. Liu eds. Beijing: Investigation of Resources and the Environment along the the Tarim River through Remote Sensing and Documentation Press.

Liu, P.X., Peng, J.F. & Chen, F.H. 2007a. Hydrological Response of *Populus euphratica* Olve. Radial Growth in Ejina Banner, Inner Mongolia. Journal of Integrative Plant Biology, 49: 150–156.

Liu, W.T., Marsh, T.L., Cheng, H. & Forney, L.J. 1997. Characterization of microbial diversity by determining terminal restriction fragment length polymorphisms of genes encoding 16S rRNA. Applied and Environmental Microbiology, 63(11): 4516–4522.

Liu, Z., Jin, Z., Li, Y., Li, T., Gu, J. & Gao, S. 2007b. Sediment phosphorus fractions and profile distribution at different vegetation growth zones in a macrophyte dominated shallow Wuliangsuhai Lake, China. Environmental Geology, 52(5): 997–1005.

Lovley, D.R. 1995. Microbial reduction of iron, manganese, and other metals. Advances in Agronomy, 54: 175–231.

Lozupone, C. & Knight, R. 2007. Global patterns in bacterial diversity. Proceedings of the National Academy of Sciences of the United States of America, 104(27): 11436–40.

Lu, H., Burbank, D.W. & Li, Y. 2009. Alluvial sequence in the north piedmont of the Chinese Tian Shan over the past 550kyr and its relationship to climate change. Palaeogeography, Palaeoclimatology, Palaeoecology, doi: 10.1016/j.palaeo.2009.11.031.

Lu, J., Zhong, S., Li, L. & Pan, S. 2007. Analysis of ecological restoration of inland desert oasis based on remote sensing. [Online]. 2007. Available from: http://www.digitwater.net/articleinfo.asp?IDD113 [Accessed: 15 December 2013].

Lüderitz, V. & Jüpner, R. 2009. Renaturierung von Fließgewässern. In: S. Zerbe & G. Wiegleb eds. Renaturierung von Ökosystemen in Mitteleuropa. Heidelberg: Spektrum Akademischer Verlag, 95–124.

Lüderitz, V., Zerbe, S., Jüpner, R. & Arevalo, J.R. 2010. Ecosystem restoration and sustainable management of rivers and wetlands – Introduction to the special issue. Forest Ecology, Landscape Research and Nature Conservation, 10: 5–6.

De Maeseneer, J.L. 1997. Constructed wetlands for sludge dewatering. In: Water Science and Technology. 1997, 279–285.

Man, R. De & Croon, F. 2007. YRD Green Pulp Partnership Frank Croon. [Online]. 2007. YRD Green Pulp Partnership. Available from: http://rdeman.nl/site/download/GreenPulpPartnershipHelsinki2007l.pdf [Accessed: 13 December 2013].

Mant, J. & Janes, M. 2006. Restoration of rivers and floodplains. In: J. van Andel & J. Aronson eds. Restoration ecology: The new frontier. Malden, Mass: Blackwell Science, 141–157.

Margesin, R. & Schinner, F. 2001. Bioremediation (natural attenuation and biostimulation) of diesel-oil-contaminated soil in an alpine glacier skiing area. Applied and Environmental Microbiology, 67(7): 3127–3133.

Margulies, M., Egholm, M., Altman, W.E., Attiya, S., Bader, J.S., Bemben, L.A., Berka, J., Braverman, M.S., Chen, Y.-J., Chen, Z., Dewell, S.B., Du, L., Fierro, J.M., Gomes, X. V, Godwin, B.C., He, W., Helgesen, S., Ho, C.H., Irzyk, G.P., Jando, S.C., Alenquer, M.L.I., Jarvie, T.P., Jirage, K.B., Kim, J.-B., Knight, J.R., Lanza, J.R., Leamon, J.H., Lefkowitz, S.M., Lei, M., Li, J., Lohman, K.L., Lu, H., Makhijani, V.B., McDade, K.E., McKenna, M.P., Myers, E.W., Nickerson, E., Nobile, J.R., Plant, R., Puc, B.P., Ronan, M.T., Roth, G.T., Sarkis, G.J., Simons, J.F., Simpson, J.W., Srinivasan, M., Tartaro, K.R., Tomasz, A., Vogt, K.A., Volkmer, G.A., Wang, S.H., Wang, Y., Weiner, M.P., Yu, P., Begley, R.F. & Rothberg, J.M. 2005. Genome sequencing in microfabricated high-density picolitre reactors. Nature, 437(7057): 376–380.

McGeoch, M.A. 1998. The selection, testing and application of terrestrial insects as bioindicators. Biological Reviews of the Cambridge Philosophical Society, 73(2): 181–201.

McGill, B.J., Enquist, B.J., Weiher, E. & Westoby, M. 2006. Rebuilding community ecology from functional traits. Trends in Ecology and Evolution, 21(4): 178–185.

Mckentry, P. 2002. Energy production biomass (part 1): Overview of biomass. Bioresource Technology, 83: 37–46.

MEA. 2005. Ecosystems and Human Well-Being: Wetlands and Water Synthesis. World Resources Institute ed. Washington, DC: Millennium Ecosystem Assessment.

Micsinai, A., Borsodi, A.K., Csengeri, V., Horváth, A., Oravecz, O., Nikolausz, M., Reskóné, M.N. & Márialigeti, K. 2003. Rhizome-associated bacterial communities of healthy and declining reed stands in Lake Velencei, Hungary. Hydrobiologia, 506–509(1–3): 707–713.

Mills, R., Perrigo, R., Brown, A., Rafty, M., Imberger, J., Wood, C., Pearce, D., Zuvela, P. & Siano, C. 2005. Sustainable Subiaco: The local scale application of the index of sustainable functionality. [Online]. 2005. Rethinking Development: Local Pathways to Global Wellbeing, The Second International Conference on Gross National Happiness. Available from: http://www.gpiatlantic.org/conference/papers/mills.htm [Accessed: 12 January 2014].

Mills, S. 2013. Sustainable management of reedbeds for conservation. In: Reed as a Renewable Resource. 2013, 14–16 February 2013: University of Greifswald.

Moch, J. 2013. Renewable Energy In China: An Overview. Climate and Energy Program, World Resources Institute, [Online]. Available from: http://www.chinafaqs.org/library/chinafaqs-renewable-energy-china-overview-0 [Accessed: 12 October 2013].

MoEP & QSIQ. 2002. Environmental quality standards for surface water. [Online]. 2002. Ministry of Environmental Protection of the People's Republic of China and State Administration for Quality Supervision and Inspection and Quarantine. Available from: http://english.mep.gov.cn/SOE/soechina1997/water/standard.htm [Accessed: 12 October 2010].

MOFCOM. 2012. Hu Jintao's Report at Eighteenth Party Congress. [Online]. 2012. Ministry of Commerce, PRC. Available from: http://english.mofcom.gov.cn/ [Accessed: 30 April 2014].

Morgan, P. & Watkinson, R.J. 1989. Hydrocarbon degradation in soils and methods for soil biotreatment. Critical Reviews in Biotechnology, 8(4): 305–333.

Mothes, F., Reiche, N., Fiedler, P., Moeder, M. & Borsdorf, H. 2010. Capability of headspace based sample preparation methods for the determination of methyl tert-butyl ether and benzene in reed (Phragmites australis) from constructed wetlands. Chemosphere, 80(4): 396–403.

Muyzer, G., Hottenträger, S., Teske, A. & Wawer, C. 1996. Denaturing gradient gel electrophoresis of PCR-amplified 16S rDNA a new molecular approach to analyse the genetic diversity of mixed microbial communities. Molecular Microbial Ecology Manual, 3: 1–23.

MWR. 2004. Pilot experiences of establishing a water-saving society in China. Ministry of Water Resources ed. Beijing: China Water Press.

NDRC & MoWR. 1998. Plan on annual allocation of water and plan on mainstream water regulation of the Yellow. National Development and Reform Commission of the People's Republic of China and Ministry of Water Resources of the People's Republic of China.

NDRC. 2006. China's National Climate Change Program. National Assessment Report on Climate Change. [Online]. 2006. Available from: http://www.china.org.cn/english/environment/213624.htm [Accessed: 18 November 2013].

Neefs, J.M., Van de Peer, Y., De Rijk, P., Chapelle, S. & De Wachter, R. 1993. Compilation of small ribosomal subunit RNA structures. Nucleic Acids Research, 21(13): 3025–3049.

Nelson, V. 2011. Introduction to Renewable Energy. A. Ghassemi ed. CRC Press, Taylor & Francis Group.

Nemergut, D.R., Robeson, M.S., Kysela, R.F., Martin, A.P., Schmidt, S.K. & Knight, R. 2008. Insights and inferences about integron evolution from genomic data. BMC genomics, 9: 261.

New Economics Foundation. 2006. Happy Planet Index. [Online]. 2006. Available from: http://www.happyplanetindex.org/ [Accessed: 4 October 2013].

Norton, B.G. 2005. Sustainability: A Philosophy of Adaptive Ecosystem Management. Chicago: University of Chicago Press.

Novikova, N.M. 2001. Ecological Basis for Botanical Diversity Conservation within the Amudarya and Syrdarya River Delta. In: S. W. Breckle, M. Veste, & W. Wucherer eds. Sustainable Land-use in Deserts. Heidelberg, Germany: Springer, 84–94.

OECD. 2013. OECD Economic Surveys: China. [Online]. 2013. Available from: http://www.oecd-ilibrary.org/economics/oecd-economic-surveys-china_20725027 [Accessed: 22 July 2013].

Ogar, N.P.P. 2003. Vegetation of river valleys. In: E. I. Rachkovskaya, E. A. Volkova, & V. N. Khramtsov eds. Botanical geography of Kazakhstan and middle Asia (Desert region). Komarov Botanical Institute of Russian Academy of Sciences. Saint Petersburg, Institute of Botany and

Phytointroduction of Ministry of Education and Science of Republic Kazakhstan. Almaty, Institute of Botany of Academy of Sciences of Republik Uzbekistan, 313–339.

Ostendorp, W. 1999. Management Impacts on Stand Structure of Lakeshore *Phragmites* Reeds. International Review of Hydrobiology, 84: 33–47.

Ott, K. 1997. Ipso Facto. Frankfurt: Suhrkamp Verlag.

Ott, K. 2008. A Modest Proposal of How to Proceed in Order to Solve the Problem of Inherent Moral Value in Nature. In: L. Westra, K. Bosselmann, & R. Westra eds. Reconciling Human Existence with Ecological Integrity. London: Earthscan, 39–60.

Ott, K. 2010. Umweltethik zur Einführung. Hamburg: Junius Verlag.

Ott, K. 2014. Institutionalizing Strong Sustainability: A Rawlsian Perspective. Sustainability, 6(2): 894–912.

Ott, K. & Döring, R. 2008. Theorie und Praxis starker Nachhaltigkeit. Marburg: Metropolis-Verlag.

Ott, K., Muraca, B. & Baatz, C. 2011. Strong Sustainability as a Frame for Sustainability Communication. In: J. Godemann & G. Michelsen eds. Sustainability Communication: Interdisciplinary Perspectives and Theoretical Foundation. Dordrecht: Springer Netherlands, 13–25.

Paavilainen, L. 1998. Modern non-wood pulp mill - process concepts and economic aspects. In: North America Nonwood Fiber Symposium Proceedings. 1998, 17–18 February 1998: Atlanta, Georgia.

Pacific Institute. 2014. Sustainable Water Management - Local to Global. [Online]. 2014. Available from: www.pacinst.org [Accessed: 30 April 2014].

Park, M. 2009. Pollutants Removal and Distribution of Microorganisms in a Reed Wetland of Shanghai. Environmental Progress, 28: 240–248.

Park, N., Vanderford, B.J., Snyder, S.A., Sarp, S., Kim, S.D. & Cho, J. 2009. Effective controls of micropollutants included in wastewater effluent using constructed wetlands under anoxic condition. Ecological Engineering, 35(3): 418–423.

Park, Y.-K., Yoo, M.L., Heo, H.S., Lee, H.W., Park, S.H., Jung, S.-C., Park, S.-S. & Seo, S.-G. 2012. Wild reed of Suncheon Bay: Potential bio-energy source. Renewable Energy, 42: 168–172.

Parr, L.B. & Mason, C.F. 2003. Long-term trends in water quality and their impact on macroinvertebrate assemblages in eutrophic lowland rivers. Water Research, 37(12): 2969–2979.

Patuzzi, F., Mimmo, T., Cesco, S., Gasparella, A. & Baratieri, M. 2013a. Common reeds (*Phragmites australis*) as sustainable energy source: Experimental and modelling analysis of torrefaction and pyrolysis processes. GCB Bioenergy, 5(4): 367–374.

Patuzzi, F., Roveda, D., Mimmo, T., Karl, J. & Baratieri, M. 2013b. A comparison between on-line and off-line tar analysis methods applied to common reed pyrolysis. Fuel, 111: 689–695.

Pei-dong, Z., Guomei, J. & Gang, W. 2007. Contribution to emission reduction of CO_2 and SO_2 by household biogas construction in rural China. Renewable and Sustainable Energy Reviews, 11(8): 1903–1912.

Perelet, R. 2007. Central Asia: Background Paper on Climate Change. [Online]. 2007. UNDP: Human Development Report 2007/2008. Available from: http://hdr.undp.org/en/reports/global/hdr2007-2008/papers/ perelet _renat.pdf [Accessed: 18 January 2014].

Perelo, L.W. 2010. Review: In situ and bioremediation of organic pollutants in aquatic sediments. Journal of Hazardous Materials, 177(1-3): 81–89.

Pérez-Harguindeguy, N., Díaz, S., Garnier, E., Lavorel, S., Poorter, H., Jaureguiberry, P., Bret-Harte, M.S., Cornwell, W.K., Craine, J.M., Gurvich, D.E., Urcelay, C., Veneklaas, E.J., Reich, P.B., Poorter, L., Wright, I.J., Ray, P., Enrico, L., Pausas, J.G., de Vos, A.C., Buchmann, N., Funes, G., Quétier, F., Hodgson, J.G., Thompson, K., Morgan, H.D., ter Steege, H., van der Heijden, M.G.A., Sack, L., Blonder, B., Poschlod, P., Vaieretti, M. V, Conti, G., Staver, A.C., Aquino, S. & Cornelissen, J.H.C. 2013. New handbook for standardised measurement of plant functional traits worldwide. Australian Journal of Botany, 61(3): 167–234.

Perrow, M.R. & Davy, A.J. 2002. Handbook of ecological restoration (Vol. 1 and 2). In: Cambridge, UK: Cambridge University Press.

Piao, S., Ciais, P., Huang, Y., Shen, Z., Peng, S., Li, J., Zhou, L., Liu, H., Ma, Y., Ding, Y., Friedlingstein, P., Liu, C., Tan, K., Yu, Y., Zhang, T. & Fang, J. 2010. The impacts of climate change on water resources and agriculture in China. Nature, 467(7311): 43–51.

Piddock, L.J. V. 2006. Multidrug-resistance efflux pumps – not just for resistance. Nature reviews. Microbiology, 4(8): 629–636.

Pilon-Smits, E. 2005. Phytoremediation. Annual Review of Plant Biology, 56: 15–39.

Po, M., Kaercher, J.D. & Nancarrow, B.E. 2003. Literature Review of Factors Influencing Public Perceptions of Water Reuse, Canberra.

Poole, K. 2005. Efflux-mediated antimicrobial resistance. The Journal of Antimicrobial Chemotherapy, 56(1): 20–51.

Pöyry. 2006. Technical Report, Modules 2–7. In: China: Technical Assistance for the Sustainable Development of the Non-wood Pulp and Paper Industry. Shanghai: Pöyry Forest Industry Co. Ltd.

Prando, D., Patuzzi, F., Pernigotto, G., Gasparella, A. & Baratieri, M. 2013. Biomass CHP Systems for Residential Applications: A Multi-Stage Modeling Approach. In: K. Kabele, M. Urban, K. Suchý, & M. Lain eds. CLIMA 2013 - 11th REHVA World Congress and the 8th International Conference on Indoor Air Quality, Ventilation and Energy Conservation in Buildings. June 2013, Prague: Society of Environmental Engineering (STP), 3309–3318.

Prashar, P., Kapoor, N. & Sachdeva, S. 2013. Rhizosphere: its structure, bacterial diversity and significance. Reviews in Environmental Science and Bio/Technology, 1705: 1–15.

Prins, M.J., Ptasinski, K.J. & Janssen, F.J.J.G. 2006. More efficient biomass gasification via torrefaction. Energy, 31(15): 3458–3470.

Qi, S. & Cai, Y. 2007. Mapping and Assessment of Degraded Land in the Heihe River Basin, Arid Northwestern China. Sensors, 7(11): 2565–2578.

Qi, S.-Z. & Luo, F. 2005. Water environmental degradation of the Heihe River Basin in arid northwestern China. Environmental Monitoring and Assessment, 108(1–3): 205–215.

Quan, W.M., Han, J.D., Shen, A.L., Ping, X.Y., Qian, P.L., Li, C.J., Shi, L.Y. & Chen, Y.Q. 2007. Uptake and distribution of N, P and heavy metals in three dominant salt marsh macrophytes from Yangtze River estuary, China. Marine Environmental Research, 64(1): 21–37.

RCW. 2005. Ramsar Convention on Wetlands of International Importance. [Online]. 2005. Ramsar Convention on Wetlands. Available from: http://www.ramsar.org/ [Accessed: 19 February 2011].

Redefining Progress. 1995. Genuine Progress Indicator. [Online]. 1995. Available from: http://rprogress.org/sustainability_indicators/genuine_progress_indicator.htm [Accessed: 20 July 2013].

Rees, W. 1992. Ecological footprints and appropriated carrying capacity: What urban economics leaves out. Environment and Urbanization, 4(2): 121–130.

Reich, P.B. & Oleksyn, J. 2004. Global patterns of plant leaf N and P in relation to temperature and latitude. Proceedings of the National Academy of Sciences of the United States of America, 101(30): 11001–11006.

Reichel, F. 2013. Bauschilf (Construction reed). In: [Online]. 2013. Available from: http://www.bauschilf.at/produkte.htm [Accessed: 30 April 2013].

Ringler, C., Cai, X., Wang, J., Ahmed, A., Xue, Y., Xu, Z., Yang, E., Jianshi, Z., Zhu, T., Cheng, L., Yongfeng, F., Xinfeng, F., Xiaowei, G. & You, L. 2010. Yellow River basin: living with scarcity. Water International, 35(5): 681–701.

Rockne, K.J., Chee-Sanford, J.C., Sanford, R.A., Hedlund, B.P., Staley, J.T. & Strand, S.E. 2000. Anaerobic naphthalene degradation by microbial pure cultures under nitrate-reducing conditions. Applied and Environmental Microbiology, 66(4): 1595–1601.

Rodewald-Rudescu, L. 1958. Schilfrohr und Fischkultur im Donaudelta (Reed and fish culture in the Danube Delta). Archiv für Hydrobiologie, 54(3): 303–339.

Rodewald-Rudescu, L. 1974. Das Schilfrohr (The reed stalk). E. Schweizerbart'sche Verlagsbuchhandlung (Nägele u. Obermiller) Stuttgart.

Rösch, C., Mergel, A. & Bothe, H. 2002. Biodiversity of denitrifying and dinitrogen-fixing bacteria in an acid forest soil. Applied and Environmental Microbiology, 68(8): 3818–3829.

Rosemond, A.D., Mulholland, P.J. & Elwood, J.W. 1993. Top-Down and Bottom-Up Control of Stream Periphyton: Effects of Nutrients and Herbivores. Ecology, 74(4): 1264–1280.

Royal Government of Bhutan. 1999. Gross National Happiness. [Online]. 1999. Available from: http://www.grossnationalhappiness.com/ [Accessed: 14 July 2013].

Rudescu, L., Niculescu, C. & Chivu, I.P. 1965. Monografia stufului din Delta Dunarii. Editura Academiei Republicii Socialiste Romania. Bucharest.

Rüger, N., Schlüter, M. & Matthies, M. 2005. A fuzzy habitat suitability index for *Populus euphratica* in the Northern Amudarya delta (Uzbekistan). Ecological Modelling, 184(2–4): 313–328.

Ruttkay, A., Tielesch, S. & Veszprémi, B. 1964. Schilfrohrwirtschaft. - Mezogazdasági Kiadó. Budapest: Landw. Verlag.

Sainty, G. 1985. Weed control and utilization of aquatic control and utilization of aquatic plants of lake Edku and barsik fish farm - Egypt. Food and Agriculture Organization of the United Nations.

Salt, D.E., Blaylock, M., Kumar, N.P., Dushenkov, V., Ensley, B.D., Chet, I. & Raskin, I. 1995. Phytoremediation: a novel strategy for the removal of toxic metals from the environment using plants. Bio/technology (Nature Publishing Company), 13(5): 468–474.

Savcor Indufor Oy. 2006. China. Technical Assistance for the Sustainable Development of the Non-Wood Pulp and Paper Industry. Impact assessment. UNEP.

SBZC. 2003. Statistics Yearbook of Zhangye City. Zhangye, China: Zhangye Press.

Schäfer, A. 1999. Schilfrohrkultur auf Niedermoor - Rentabilität des Anbaus und der Ernte von *Phragmites australis*. Archiv für Naturschutz and Landschaftsforschung, 38: 193–216.

Schelwald-van der Kley, L. & Reijerkerk, L. 2014. Water: A way of life. [Online]. 2014. Available from: http://www.waterandculture.org [Accessed: 9 April 2014].

Schuster, J. 1985. Schilfverwertung - Erntestudie. In: Grosina: AGN-Forschungsprogramm und seine Umsetzung im Raum. 586–609.

Schutter, M., Sandeno, J. & Dick, R. 2001. Seasonal, soil type, and alternative management influences on microbial communities of vegetable cropping systems. Biology and Fertility of Soils, 34(6): 397–410.

Shang, S., Cao, Y., Guang, R., Wang, Z., Zang, C. & Li, Y. 2011. Spatial temporal expansion of reed marsh in the Wuliangsuhai wetland in Inner Mongolia. In: World Environmental and Water Resources Congress 2011, 4840–4848.

Shi, B., Ma, J., Wang, K., Gong, J., Zhang, C. & Liu, W. 2010. Effects of atmospheric elevated temperature on the growth, reproduction and biomass allocation of reclamation *Phragmites australis* in East Beach of Chongming Island. Resources and Environment in the Yangtze Basin, 19(4): 383–388.

Simonet, P., Navarro, E., Rouvier, C., Reddell, P., Zimpfer, J., Dommergues, Y., Bardin, R., Combarro, P., Hamelin, J., Domenach, A.M., Gourbière, F., Prin, Y., Dawson, J.O. & Normand, P. 1999. Co-evolution between *Frankia* populations and host plants in the family Casuarinaceae and consequent patterns of global dispersal. Environmental Microbiology, 1(6): 525–533.

Singer, R.S., Ward, M.P. & Maldonado, G. 2006. Can landscape ecology untangle the complexity of antibiotic resistance? Nature reviews. Microbiology, 4(12): 943–952.

Sluis, T. Van Der, Rii, O., Poppens, R. & Lesschen, J.P. 2013. Reed resources in Poltava Oblast , Ukraine: biodiversity conservation and bioenergy production. In: Reed as a Renewable Resource. 2013, 14–16 February 2013: Greifswald, Germany.

Smalla, K., Wieland, G., Buchner, A., Zock, A., Parzy, J., Kaiser, S., Roskot, N., Heuer, H. & Berg, G. 2001. Bulk and rhizosphere soil bacterial communities studied by denaturing gradient gel electrophoresis: plant-dependent enrichment and seasonal shifts revealed. Applied and Environmental Microbiology, 67(10): 4742–4751.

Smith, S.D., Devitt, D.A., Sala, A., Cleverly, J.R. & Busch, D.E. 1998. Water relations of riparian plants from warm desert regions. Wetlands, 18(4): 687–696.

Smole, M.S., Hribernik, S., Kleinschek, K.S. & Kreže, T. 2013. Plant fibres for textile and technical applications. Advances in Agrophyical Research, 369–397.

Snow, C.P. 1959. The Two Cultures. Canto 1993. London: Cambridge University Press.

Snow, C.P. 1990. The Two Cultures. Leonardo, 23(2): 169–173.

Sogin, M.L., Morrison, H.G., Huber, J.A., Mark Welch, D., Huse, S.M., Neal, P.R., Arrieta, J.M. & Herndl, G.J. 2006. Microbial diversity in the deep sea and the underexplored "rare biosphere". Proceedings of the National Academy of Sciences of the United States of America, 103(32): 12115–12120.

Song, Y.D., Fan, Z.L., Lei, Z.D. & Zhang, F.W. 2000. Research on Water Resources and Ecology of Tarim River, China (Zhongguo Talim He Shui Ziyuan yu Shangtai Wenti Yanjiu). Urumqi: Xinjiang Peoples Press (Xinjiang Renmin Chubanshe).

Southichak, B., Nakano, K., Nomura, M., Chiba, N. & Nishimura, O. 2006. *Phragmites australis*: A novel biosorbent for the removal of heavy metals from aqueous solution. Water Research, 40(12): 2295–2302.

Sowers, K.R. & May, H.D. 2013. In situ treatment of PCBs by anaerobic microbial dechlorination in aquatic sediment: Are we there yet? Current Opinion in Biotechnology, 24(3): 482–488.

State Forestry Bureau. 2014. National Wetland Inventory Survey (2009–2013). [Online]. 2014. The Nature Conservancy Chinese Department. Available from: http://www.shidi.org/ [Accessed: 25 April 2014].

Stenman, H. 2008. Reed construction on the Baltic Sea region. Reports from Turku University of Applied Sciences.

Stepanauskas, R., Glenn, T.C., Jagoe, C.H., Tuckfield, R.C., Lindell, A.H., King, C.J. & McArthur, J. V. 2006. Coselection for microbial resistance to metals and antibiotics in freshwater microcosms. Environmental Microbiology, 8(9): 1510–1514.

Strezov, V., Evans, T.J. & Hayman, C. 2008. Thermal conversion of elephant grass (*Pennisetum purpureum* Schum) to bio-gas, bio-oil and charcoal. Bioresource Technology, 99(17): 8394–8399.

Su, N., Zhang, J. & Wang, Y. 1994. The pollution and evaluation of Cd in soil in Hujian Province. Acta Hujian Agriculture University, 23: 434–439.

Sukhova, G. V. & Gladyshev, A.I. 1980. Ecological and anatomical characteristics of Tugai vegetation in the Amu Darya USSR bottom land. Izvestiya Akademii Nauk Turkmenskoi Ssr Seriya Biologicheskikh Nauk.

Sun, M.Y., Dafforn, K.A., Brown, M. V. & Johnston, E.L. 2012. Bacterial communities are sensitive indicators of contaminant stress. Marine Pollution Bulletin, 64(5): 1029–1038.

Sutcu, H. 2008. Pyrolysis of *Phragmites Australis* and characterization of liquid and solid products. Journal of Industrial and Engineering Chemistry, 14(5): 573–577.

Taisan. 2009. Suitability of Using *Phragmites australis* and *Tamarix aphylla* as Vegetation Filters in Industrial Areas. American Journal of Environmental Sciences, 5(740): 747.

TEEB. 2010. The Economics of Ecosystems and Biodiversity: Ecological and Economic Foundations. Earthscan ed. London and Washington, DC.

Temperton, V.M. & Kirr, K. 2004. Order of arrival and availability of safe sites: An example of their importance for plant community assembly in stressed ecosystems. V. M. Temperton, R. J. Hobbs, T. Nuttle, & S. Halle eds. Washington, DC: Island Press.

Thevs, N. 2007. Ecology, Spatial Distribution, and Utilisation of the Tugai Vegetation at the Middle Reaches of the Tarim River, Xinjiang, China. Göttingen: Cuvillier.

Thevs, N., Buras, A., Zerbe, S., Kuhnel, E., Abdusalih, N. & Ovezberdiyeva, A. 2012. Structure and wood biomass of near-natural floodplain forests along the Central Asian rivers Tarim and Amu Darya. Forestry, 85(2): 193–202.

Thevs, N., Ovezmuradov, K., Zanjani, L.V. & Zerbe, S. 2014. Water consumption of agriculture and natural ecosystems at the Amu Darya in Lebap Province, Turkmenistan. Environmental Earth Sciences. doi: 10.1007/s12665-014-3084-1.

Thevs, N., Peng, H., Rozi, A., Zerbe, S. & Abdusalih, N. (2015). Water allocation and water consumption of irrigated agriculture and natural vegetation in the Aksu-Tarim river basin, Xinjiang, China. Journal of Arid Environments, 112, Part A, 87–97. doi:10.1016/j.jaridenv.2014.05.028

Thevs, N., Rozi, A., Kubal, C. & Abdusalih, N. 2013. Water consumption of agriculture and natural ecosystems along the Tarim River, China. Geo-Öko, 34: 50–76.

Thevs, N., Zerbe, S., Gahlert, E., Mijit, M. & Succow, M. 2007. Productivity of reed (*Phragmites australis* Trin. ex Steud.) in continental-arid NW China in relation to soil, groundwater, and land-use. Journal of Applied Botany and Food Quality-Angewandte Botanik, 81(1): 62–68.

Thevs, N., Zerbe, S., Peper, J. & Succow, M. 2008a. Vegetation and vegetation dynamics in the Tarim River floodplain of continental-arid Xinjiang, NW China. Phytocoenologia, 38(1): 65–84.

Thevs, N., Zerbe, S., Schnittler, M., Abdusalih, N. & Succow, M. 2008b. Structure, reproduction and flood-induced dynamics of riparian Tugai forests at the Tarim River in Xinjiang, NW China. Forestry, 81(1): 45–57.

Thevs, N., Zerbe, S., Wucherer, W., Buras, A. & Bo, T. 2011. Ecology and utilization of salt-tolerant plants in the river basins of Central Asia. Ecological Questions, 14: 19–20.

Thomas, F.M., Foetzki, A., Arndt, S.K., Bruelheide, H., Gries, D., Li, X., Zeng, F., Zhang, X. & Runge, M. 2006. Water use by perennial plants in the transition zone between river oasis and desert in NW China. Basic and Applied Ecology, 7(3): 253–267.

Tian, Z., Zheng, B., Liu, M. & Zhang, Z. 2009. *Phragmites australis* and *Typha orientalis* in removal of pollutant in Taihu Lake, China. Journal of Environmental Sciences, 21(4): 440–446.

Timmermann, T., Joosten, H. & Succow, M. 2009. Restaurierung von Mooren. In: S. Zerbe & G. Wiegleb eds. Renaturierung von Ökosystemen in Mitteleuropa. Spektrum Akademischer Verlag, Heidelberg, 55–93.

Townend, J. 2002. Practical statistics for environmental and biological scientists. West Sussex, England: John Wiley & Sons, Ltd.

Treshkin, S.Y. 2001. The Tugai Forests of Floodplain of the Amudarya River: Ecology, Dynamics and their Conservation. In: S. W. Breckle, M. Veste, & W. Wucherer eds. Sustainable Land-use in Deserts. Heidelberg, Germany: Springer, 95–102.

Tutt, M. & Olt, J. 2011. Suitability of various plant species for bioethanol production. Agronomy Research Biosystem Engineering Special Issue, (1): 261–267.

Udvardi, M. & Poole, P.S. 2013. Transport and metabolism in legume-rhizobia symbioses. Annual Review of Plant Biology, 64: 781–805.

UN. 1992. Convention on Biological Diversity. [Online]. 1992. United Nations. Available from: http://www.cbd.int/doc/legal/cbd-en.pdf [Accessed: 19 February 2011].

UN. 1987. Our Common Future: Report of the World Commission on Environment and Development. General Assembly Resolution 42/187, United Nations.

UN. 2000. United Nations Millennium Development Goals. [Online]. 2000. Available from: http://www.un.org/millenniumgoals/ [Accessed: 24 October 2013].

UNDP. 2012. Asia-Pacific Human Development Report 2012 – One Planet to Share: Sustaining Human Progress in a Changing Climate. London: Routledge.

UNDP. 2010. Human Development Index: United Nations Development Programme (UNDP). [Online]. 2010. Available from: http://hdr.undp.org/en/statistics/hdi/ [Accessed: 17 July 2013].

UNEP. 2001. The Mesopotamian Marshlands: Demise of an Ecosystem. Nairobi, Kenya.

UNESCO. 2006. Transboundary aquifers in Asia with special emphasis to China. [Online]. 2006. UNESDOC Database. Available from: http://unesdoc.unesco.org/images/0014/001483/148390e.pdf [Accessed: 27 March 2014].

UNESCO. 2009. The United Nations World Water Development Report 3: Water in a Changing World. In: Earthscan ed. World Water Assessment Programme. 349.

UNESCO. 2012. The United Nations World Water Development Report 4: Managing Water Under Uncertainty and Risk. In: Earthscan ed. World Water Assessment Programme. 407.

UNESCO. 2014. Karez Wells: Tentative List. [Online]. 2014. World Heritage List Nominations. Available from: http://whc.unesco.org/en/tentativelists/5347/ [Accessed: 9 April 2014].

Unger-Shayesteh, K., Vorogushyn, S., Farinotti, D., Gafurov, A., Duethmann, D., Mandychev, A. & Merz, B. 2013. What do we know about past changes in the water cycle of Central Asian headwaters? A review. Global and Planetary Change, 110: 4–25.

Urbanska, K.M., Hurka, H., Landolt, E., Neuffer, B. & Mummenhoff, K. 1997. Hybridization and evolution in *Cardamine* (Brassicaceae) at Urnerboden, Central Switzerland: Biosystematic and molecular evidence.

USDA. 2012. China - Peoples Republic of Cotton and Products Annual. [Online]. 2012. GAIN Report Number: CH12031. Available from: http://www.thefarmsite.com/reports/contents/chinacotmay12.pdf [Accessed: 5 December 2013].

USGS. 2011. Map of Wuliangsuhai Lake: Border of Wuliangsuhai Lake drawn from TM image acquired August 2011. [Online]. 2011. United States Geological Survey. Available from: www.usgs.gov [Accessed: 12 November 2011].

Valverde, A., Velázquez, E., Gutiérrez, C., Cervantes, E., Ventosa, A. & Igual, J.-M. 2003. Herbaspirillum lusitanum sp. nov., a novel nitrogen-fixing bacterium associated with root nodules of *Phaseolus vulgaris*. International Journal of Systematic and Evolutionary Microbiology, 53(Pt6): 1979–1983.

VanAndel, J. & Aronson, J. 2006. Restoration ecology. The new frontier. Oxford: Blackwell Publishers.

Villegas-Navarro, A., Romero González, M.C., Rosas López, E., Domínguez Aguilar, R. & Sachetin Marçal, W. 1999. Evaluation of *Daphnia magna* as an indicator of toxicity and treatment efficacy of textile wastewaters. Environment International, 25(5): 619–624.

Vladár, P., Rusznyák, A., Márialigeti, K. & Borsodi, A.K. 2008. Diversity of sulfate-reducing bacteria inhabiting the rhizosphere of *Phragmites australis* in Lake Velencei (Hungary) revealed by a combined cultivation-based and molecular approach. Microbial Ecology, 56(1): 64–75.

Vogel, T.M. 1996. Bioaugmentation as a soil bioremediation approach. Current Opinion in Biotechnology, 7(3): 311–316.

Vymazal, J., Kröpfelová, L., Švehla, J., Chrastný, V. & Štíchová, J. 2009. Trace elements in *Phragmites australis* growing in constructed wetlands for treatment of municipal wastewater. Ecological Engineering, 35(2): 303–309.

Vymazal, J., Švehla, J., Kröpfelová, L., Němcová, J. & Suchý, V. 2010. Heavy metals in sediments from constructed wetlands treating municipal wastewater. Biogeochemistry, 101: 335–356.

Wang, G., Cheng, G. & Shen, Y. 2002. Eco-environmental variations and control strategies in Heihe Corridor region in the past 50 years. Journal of Natural Resource, 17(1): 78–86.

Wang, G., Liu, J. & Tang, J. 2004. The long-term nutrient accumulation with respect to anthropogenic impacts in the sediments from two freshwater marshes (Xianghai Wetlands , Northeast China). Water Research, 38: 4462–4474.

Wang, S., Li, J. & Shi, S. 2001. Geological disease caused by ecological environment: An example of a cancer-inflicted village in Shanxi Province. Environmental Protection, 5: 42–43.

Wang, S.J., Chen, B.H. & Li, H.Q. 1996. Euphrates Poplar Forest. Beijing: China Environmental Science Press.

Wang, Y. & Liang, S. 2013. Carbon dioxide mitigation target of China in 2020 and key economic sectors. Energy Policy, 58: 90–96.

Wayman, M. 1973. Guide for planning pulp and paper enterprises. Food and Agriculture Organization Forestry and Forest Products Studies No. 18.

Weis, J.S. & Weis, P. 2004. Metal uptake, transport and release by wetland plants: implications for phytoremediation and restoration. Environment International, 30(5): 685–700.

Westermann, J., Zerbe, S. & Eckstein, D. 2008. Age structure and growth of degraded *Populus euphratica* floodplain forests in North-west China and perspectives for their recovery. Journal of integrative plant biology, 50: 536–546.

White, G. 2009. The future of reedbed management. Information and Advice Note, 7 July 2009, RSPB.

Wiehle, M., Eusemann, P., Thevs, N. & Schnittler, M. 2009. Root suckering patterns in *Populus euphratica* (Euphrates poplar, Salicaceae). Trees, 23(5): 991–1001.

WLI. 2012. Wetland Link International. [Online]. 2012. The Wetlands of Central Asia and the Ramsar Convention. Available from: http://wli.wwt.org.uk/2012/10/members/asia/asia-news/international-conference-the-wetlands-of-central-asia-and-the-ramsar-convention/ [Accessed: 12 April 2013].

Woese, C.R. 1987. Bacterial Evolution. Microbiology and Molecular Biology Reviews, 51: 221–271.

Wright, I.J., Reich, P.B., Westoby, M., Ackerly, D.D., Baruch, Z., Bongers, F., Cavender-Bares, J., Chapin, T., Cornelissen, J.H.C., Diemer, M., Flexas, J., Garnier, E., Groom, P.K., Gulias, J., Hikosaka, K., Lamont, B.B., Lee, T., Lee, W., Lusk, C., Midgley, J.J., Navas, M.-L., Niinemets, U., Oleksyn, J., Osada, N., Poorter, H., Poot, P., Prior, L., Pyankov, V.I., Roumet, C., Thomas, S.C., Tjoelker, M.G., Veneklaas, E.J. & Villar, R. 2004. The worldwide leaf economics spectrum. Nature, 428(6985): 821–827.

Wu, B. & Ci, L.J. 2002. Landscape change and desertification development in the Mu Us Sandland, Northern China. Journal of Arid Environments, 50(3): 429–444.

Wu, J., Ma, L., Yu, H., Zeng, H., Liu, W. & Abuduwaili, J. 2013. Sediment geochemical records of environmental change in Lake Wuliangsu, Yellow River Basin, north China. Journal of Paleolimnology, 50(2): 245–255.

Wu, N., Qiao, M., Zhang, B., Cheng, W.-D. & Zhu, Y.-G. 2010. Abundance and diversity of tetracycline resistance genes in soils adjacent to representative swine feedlots in China. Environmental Science & Technology, 44(18): 6933–6939.

Wu, W. 2012. Why should attention be paid to Wuliangsuhai? [Online]. 2012. People's Daily. Available from: http://www.qstheory.cn/st/stsp/201206/t20120614_164046.htm [Accessed: 20 December 2012].

Wulate Qianqi Government. 2012. Statistics Office. Wulate Qianqi, China.

WWF and IUCN. 1996. Forests for Life - The WWF/IUCN forest policy book. Surrey: WWF-UK.

WWF. 2005. Living Planet Index. [Online]. 2005. Available from: http://wwf.panda.org/ [Accessed: 17 July 2013].

Xia, J., Zuo, Q. & Pang, J. 2001. Enlightenment on sustainable management of water resources from past practices in the Bositeng Lake basin, Xinjiang, China. In: Regional Management of Water Resources. 2001, IAHS Scientifits Assembly at Maastrich, 41–48.

Xiao, D.N. & Li, X.Z. 2004. Ecological and environmental function of wetland landscape in the Liaohe Delta. In: Devlopments in Ecosystems. Elsevier B.V., 35–46.

Xinjiang Linkeyuan Zaolin Zhisha Yanjiusuo. 1989. Huyanglin Gengxin Fuzhuang Jishu Yanjiu. Urumqi: Xinjiang Linkeyuan Zaolin Zhisha Yanjiusuo.

Yamian, Z., Yifei, J., Shengwu, J., Qing, Z., Duoduo, F., Yumin, G., Guangchun, L., Zhang, Y., Jia, Y., Jiao, S., Zeng, Q., Feng, D., Guo, Y. & Lei, G. 2012. Wuliangsuhai Wetlands: A critical Habitat for Migratory Water Birds. Journal of Resources and Ecology, 3(4): 316–323.

Yang, Q., Yin, X., Wu, C., Wu, S. & Guo, D. 2012. Thermogravimetric-Fourier transform infrared spectrometric analysis of CO_2 gasification of reed (*Phragmites australis*) kraft black liquor. Bioresource Technology, 107: 512–6.

Ye, S., Brix, H. & Sun, D. 2013. Large-scale management of common reed, *Phragmites australis*, for paper production: a case study from the Liaohe River Delta, China. In: International Conference "Reed as a Renewable Resource" 14–16 February 2013. 2013, Greifswald, Germany: Institute for Botany and Landscape Ecology, University of Greifswald.

Ye, Z., Baker, A.J.M., Wong, M.H. & Willis, A.J. 1997. Zinc, Lead and Cadmium Tolerance, Uptake and Accumulation by the Common Reed, *Phragmites australis* (Cav.) Trin. ex Steudel. Annals of Botany, 80(3): 363–370.

Yimit, H., He, L., Halik, W. & Giese, E. 2006. Water resource development and its environmental effects in the Tarim River floodplain. In: T. Hoppe, B. Kleinschmit, B. Roberts, N. Thevs, & Ü. Halik eds. Watershed and

Floodplain Management along the Tarim River in China's Arid Northwest. Aachen: Shaker, 91–107.

Yu, J., Vodyanik, M.A., Smuga-Otto, K., Antosiewicz-Bourget, J., Frane, J.L., Tian, S., Nie, J., Jonsdottir, G.A., Ruotti, V., Stewart, R., Slukvin, I.I. & Thomson, J.A. 2007. Induced pluripotent stem cell lines derived from human somatic cells. Science, 318(5858): 1917–1920.

Yuan, J., Xu, Y., Zhang, X., Hu, Z. & Xu, M. 2014. China's 2020 clean energy target: Consistency, pathways and policy implications. Energy Policy, 65: 692–700.

Van der Zanden, A.M.M. & Cook, T.W. 2011. Sustainable Landscape Management: Design, Construction, and Maintenance. Wiley.

Zeng, Q., Zhang, Y., Jia, Y., Jiao, S., Feng, D., Bridgewater, P. & Lei, G. 2012. Zoning for management in wetland nature reserves: A case study using Wuliangsuhai Nature Reserve, China. SpringerPlus, 1(1): 23.

Zerbe, S., Halik, Ü. & Küchler, J. 2006. Urban greening in the oases of continental arid Southern Xinjiang (NW China) – an interdisciplinary approach. Die Erde, 136: 245–266.

Zerbe, S. & Thevs, N. 2011. Restoring Central Asian floodplain ecosystems as natural capital and cultural heritage in a continental desert environment. In: S. K. Hong, J. Wu, J. E. Kim, & N. Nakagoshi eds. Landscape Ecology in Asian Cultures, Ecological Research Monographs. Springer.

Zerbe, S. & Wiegleb, G. 2009. Renaturierung von Ökosystemen in Mitteleuropa. Spektrum Akademischer Verlag, Heidelberg.

Zhang, H., Shen, W.S., Wang, Y.S. & Zou, C.X. 2005. Study on grassland grazing capacity in the Heihe River Basin. Journal of Natural Resources, 20: 514–521.

Zhang, J. 2012. Länderreport Strohenergie in China. In: A. Schütte & A. Vetter eds. Gülzower Fachgespräche - 2. Fachtagung Strohenergie. March 2012, Berlin: Fachagentur Nachwachsende Rohstoffe e.V. (FNR), 26–27.

Zhang, J.J. & Smith, K.R. 2007. Household air pollution from coal and biomass fuels in China: Measurements, health impacts, and interventions. Environmental Health Perspectives, 115(6): 848–855.

Zhang, K., Chang, J., Guan, Y., Chen, H., Yang, Y. & Jiang, J. 2013a. Lignocellulosic biomass gasification technology in China. Renewable Energy, 49: 175–184.

Zhang, N. 1999. Advance of the research on heavy metals in soil-plant system. Advance in Environmental Science, 7: 30–33.

Zhang, W., Wu, X., Liu, G., Chen, T., Zhang, G., Dong, Z., Yang, X. & Hu, P. 2013b. Pyrosequencing Reveals Bacterial Diversity in the Rhizosphere of Three *Phragmites australis* Ecotypes. Geomicrobiology Journal, 30(7): 593–599.

Zhang, Y., Yu, J., Wang, P. & Fu, G. 2011. Vegetation responses to integrated water management in the Ejina basin, northwest China. Hydrological Processes, 25: 3448–3461.

Zhao, H., Yan, H., Zhang, C., Liu, X., Xue, Y., Qiao, Y., Tian, Y. & Qin, S. 2011. Pyrolytic Characteristics and Kinetics of *Phragmites australis*. Evidence-Based Complementary and Alternative Medicine, 2011(9): 1–6.

Zhou, X., Wang, F., Hu, H., Yang, L., Guo, P. & Xiao, B. 2011. Assessment of sustainable biomass resource for energy use in China. Biomass and Bioenergy, 35(1): 1–11.

Zhou, Z., Wu, W., Chen, Q. & Chen, S. 2008. Study on sustainable development of rural household energy in northern China. Renewable and Sustainable Energy Reviews, 12(8): 2227–2239.

Zhu, Y., Ren, L., Skaggs, T.H., Lue, H., Yu, Z., Wu, Y. & Fang, X. 2009. Simulation of *Populus euphratica* root uptake of groundwater in an arid woodland of the Ejina Basin, China. Hydrological Processes, 23(17): 2460–2469.

Zhu, Z., Jiang, C., Zhong, M. & Huafu, W. 1998. Status, trends and prospects for non-wood and recycled fibre in China. Asia-Pacific Forestry Sector Outlook Study Working Paper Series: FAO.

Ziegler, R. & Ott, K. 2011. The quality of sustainability science: A philosophical perspective. Sustainability: Science, Practice, and Policy, 7(1): 31–44.

The SuWaRest Project – Photographic documentation

Photograph 1 – Polluted water channel with growth of *Phragmites australis* in the urbanised area of Zhangye, Gansu Province *(L. Li, 9 August 2011)*.

Photograph 2 – Tugai forest in the reach of flood events with root suckers, Ejina oasis, Inner Mongolia *(J. F. Köbbing, 25 September 2011)*.

Photograph 3 – *Populus euphratica* tree at the desert margin with deadwood, Black City, Inner Mongolia (J. F. Köbbing, 27 September 2011).

Photograph 4 – Old Tugai forest along the eastern branch of the Heihe in Ejina (J. F. Köbbing, 26 September 2011).

Photograph 5 – Tugai forest stretched along the eastern branch of the Heihe in Ejina *(J. F. Köbbing, 26 September 2011).*

Photograph 6 – Summer reed found throughout the wetland areas of Wuliangsuhai Lake, Inner Mongolia *(L. Li, 5 August 2011).*

Photograph 7 – Winter reed along the outskirts of Wuliangsuhai Lake, Inner Mongolia
(L. Li, 16 February 2011).

a b

Photograph 8 – Traditional heating devices in rural China, Inner Mongolia:
[a] coal stove and [b] Kang, coupled stove and heated bed using *Populus euphratica* as deadwood
(F. Patuzzi, 28 September 2011).

Photograph 9 – Bundled reed as fodder plant at Wuliangsuhai Lake, Inner Mongolia
(J. F. Köbbing, 3 September 2012).

Photograph 10 – Reed beds grazed by goat at Wuliangsuhai Lake, Inner Mongolia
(J. F. Köbbing, 3 September 2012).

Photograph 11 – Bora mats for drying fruits, Luntai County, Xinjiang
(R. Aihemaitijiang, 5 January 2007).

Photograph 12 – Traditional manual weaving of bora mats in Xinjiang
(R. Aihemaitijiang, 12 May 2013).

Photograph 13 – Production of Yuban reed panels in Yanji County, Xinjiang
(R. Aihemaitijiang, 1 January 2013).

Photograph 14 – Loading of Yuban reed panels, Yanji County, Xinjiang
(R. Aihemaitijiang, 12 May 2013).

Photograph 15 – Weaving loom for reed mats at Wuliangsuhai Lake, Inner Mongolia *(J. F. Köbbing, 10 September 2012)*.

Photograph 16 – Bundled and stored finished reed mats at Wuliangsuhai Lake, Inner Mongolia *(J. F. Köbbing, 10 September 2012)*.

Photograph 17 – Loading reed onto a pressing machine for bales at Wuliangsuhai Lake, Inner Mongolia *(J. F. Köbbing, 3 September 2012)*.

Photograph 18 – Reed pressing machines for bales at Wuliangsuhai Lake, Inner Mongolia *(J. F. Köbbing, 3 September 2012)*.

Photograph 19 – Loading of reed trucks at Wuliangsuhai, Inner Mongolia *(J. F. Köbbing, 4 September 2012)*.

Photograph 20 – Weighing and transport of reed trucks before leaving Wuliangsuhai Lake, Inner Mongolia *(J. F. Köbbing, 4 September 2012)*.

Photograph 21 – Reed trucks transporting bales from local sellers around Wuliangsuhai Lake, Inner Mongolia *(J. F. Köbbing, 11 September 2012)*.

Photograph 22 – Isolated mountainous terrain of Xiaoshetai County in northern Urat Front Banner, Inner Mongolia *(G. T. Cirella, 8 October 2013)*.

Photograph 23 – Wuliangsuhai Yuchang Batou village located in the east of Erdenbulage County, Inner Mongolia *(G. T. Cirella, 30 September 2013)*.

Photograph 24 – Gongtian village, Xixiaozhao County in western Urat Front Banner is a part of the southern transportation route of the Bayannur *(G. T. Cirella, 3 October 2013)*.

www.ingramcontent.com/pod-product-compliance
Lightning Source LLC
LaVergne TN
LVHW082128200926
848484LV00008B/2023